# DOING NUTRITION DIFFER

# Critical Food Studies

*Series Editor*
*Michael K. Goodman, Kings College London, UK*

The study of food has seldom been more pressing or prescient. From the intensifying globalization of food, a world-wide food crisis and the continuing inequalities of its production and consumption, to food's exploding media presence, and its growing re-connections to places and people through 'alternative food movements', this series promotes critical explorations of contemporary food cultures and politics. Building on previous but disparate scholarship, its overall aims are to develop innovative and theoretical lenses and empirical material in order to contribute to – but also begin to more fully delineate – the confines and confluences of an agenda of critical food research and writing.

Of particular concern are original theoretical and empirical treatments of the materializations of food politics, meanings and representations, the shifting political economies and ecologies of food production and consumption and the growing transgressions between alternative and corporatist food networks.

# Doing Nutrition Differently
## Critical Approaches to Diet and Dietary Intervention

*Edited by*

ALLISON HAYES-CONROY
*Temple University, USA*

JESSICA HAYES-CONROY
*Hobart and William Smith Colleges, USA*

LONDON AND NEW YORK

First published 2013 by Ashgate Publishing

2 Park Square, Milton Park, Abingdon, Oxon OX14 4RN
711 Third Avenue, New York, NY 10017, USA

*Routledge is an imprint of the Taylor & Francis Group, an informa business*

First issued in paperback 2016

**British Library Cataloguing in Publication Data**
A catalogue record for this book is available from the British Library

**The Library of Congress has cataloged the printed edition as follows:**
Doing nutrition differently : critical approaches to diet and dietary intervention /
[edited] by Allison Hayes-Conroy and Jessica Hayes-Conroy.
    pages cm. -- (Critical food studies)
 Includes bibliographical references and index.
 ISBN 978-1-4094-3479-5 (hardback)
1. Nutrition. 2. Women--Nutrition--Psychological aspects. 3. Nutritionally induced diseases--Prevention. 4. Health behavior. I. Hayes-Conroy, Allison, 1981-, editor of compilation. II. Hayes-Conroy, Jessica, editor of compilation.

 RA784.D63 2013
 613.2--dc23

                                                                        2013020282

ISBN 978-1-4094-3479-5 (hbk)
ISBN 978-1-138-25284-4 (pbk)

# Contents

*Thematic tabs:*

Access

Emotion

Science

Structure

*Thematic tabs:*

Access

Body

Colonial

Emotion

Nature

Race

Women

# List of Figures

# Notes on Contributors

**Alison Hope Alkon** is an assistant professor in the Department of Sociology at the University of the Pacific, in Stockton, CA. She studies environmental and social justice, social inequality, and agri-food systems. She is the co-editor of the recently published volume, *Cultivating Food Justice: Race, Class and Sustainability* (The MIT Press), and is author of a number of articles on food justice, food activism and race.

**Tamara Beauboeuf-Lafontant** is Associate Professor of Women's Studies at DePauw University. Her work on constructions of good womanhood and their impact on women's physical and psychological wellness has been published in *Meridians, Gender & Society,* and *Qualitative Sociology*. She is author of the recently published, *Behind the Mask of the Strong Black Woman: Voice and the Embodiment of a Costly Performance* (Temple University Press 2009).

**Kirsten Valentine Cadieux** is a researcher and lecturer in Geography and Sociology at the University of Minnesota. Her work focuses on the ways that urban and rural environmental ideologies and anxieties relate to land use practices—and on the ways that groups negotiate conflicts and contradictions in their aspirations for food systems, residential landscapes, and environmental planning. She has recently edited, with Patrick Hurley, a special issue of *GeoJournal* on amenity migration and exurban and rural land uses, and, with Laura Taylor, the collection *Landscape and the Ideology of Nature in Exurbia: Green Sprawl*.

**Sally Fallon Morell** is author of the best-selling cookbook *Nourishing Traditions* and founding president of the Weston A. Price Foundation. She is a well-known advocate for a return to nutrient-dense foods including animal fats, raw milk and cod liver oil.

**David V. II Fazzino** is a cultural anthropologist trained in law and agroecology. His research interests include food security, sustainable agriculture, local agriculture, alternative food institutions and food deserts. He is currently an assistant professor of Anthropology at the University of Alaska Fairbanks.

**Laura Frank** is a tenured professor and Chair of the Nutrition and Dietetics Department at Immaculata University, where she has taught as a professor for 15 years. She is also a registered Dietitian/PA Licensed Dietitian-Nutritionist, PA Licensed Psychologist, and Certified Rehabilitation Counselor, with more than 25 years of experience in health care/education, teaching, and counseling. In writing her chapter she draws on her professional training and practice experience, and many years of teaching Health and Nutrition Counseling to health care providers, primarily dietitians and dietitians-in-training.

**A. Breeze Harper** is a PhD Candidate at the University of California-Davis in the geography graduate group. Her dissertation work explores how race, whiteness, and decolonial politics operate within U.S. vegan consumer philosophies. She is also the editor of the anthology *Sistah Vegan: Black Female Vegans Speak on Food, Identity, Health and Society* (Lantern Books 2010).

**Edmund M. Harris** is a PhD candidate in Geography at Clark University. His research focuses on the spatial imaginaries of place, scale and the local used to envision alternative, more sustainable agriculture and food systems. His previous research explored the politics of localism in emergent alternative food networks in Scotland. He is author of several articles including "Neoliberal subjectivities or a politics of the possible? Reading for difference in alternative food networks," published in Area (41: 1, 55-63, 2009), for which he received the 2009 Area prize for New Research in Geography. For more information, see edmundharris.com.

**Allison Hayes-Conroy** is an assistant professor in the department of Geography and Urban Studies at Temple University. She has authored two books on the culture and politics of food and agriculture, as well as a number of papers on the visceral politics of food. Her current research centers on novel approaches to food security and non-violence in Medellin, Colombia.

**Jessica Hayes-Conroy** has a PhD in geography and women's studies from Penn State University. She recently served a Mellon Post-Doctoral Fellow in Environmental Studies and Women's Studies at Wheaton College, and is now an assistant professor of Women's Studies at Hobart and Williams Smith Colleges. She has authored papers on alternative food, visceral geography, and political ecology. Her current research centers on critical perspectives of nutrition intervention.

**Philip A. Loring** is a human ecologist with interests in food systems, environmental change, and environmental justice. His current research involves coastal communities in Alaska, with a focus on the role of local seafood in community food security, and on how state-based calculations and certifications of 'sustainability' for Alaska fisheries are obstructing meaningful social change.

**Laura Newcomer** is a writer, editor, and ecology educator with Bachelor's degrees in English and Geography. Until 2011, she was employed in editorial roles first with the National Geographic Society and then with the American Association for Justice in Washington, DC. She then entered the world of ecology education, working for Tanglewood 4-H Camp and Learning Center and the Ferry Beach

Ecology School in Maine. Currently, she works as a freelance Researcher and as the Happiness Editor at Greatist, an online resource for all things pertaining to health and wellness. Laura was anorexic for eight years and rooted her recovery in creative expression. She maintains the website www.lightinfinityexpress.blogspot. com and writes sporadically for a variety of publications.

**Chris Rodriguez** is the founder of Decolonial Food For Thought, an ethnonutritional centered community health project. His work as a blogger, community chef, educator, and activist strongly embodies principles gained from his dedicated active involvement with La Sexta. Rodriguez is highly vocal and organizes on the critical issues of Indigenous food autonomy, ethnoecology, the cultivation of kitchenspaces, and food as a healing and reindigenizing memory source, enacting responsibility and healthy eating practices.

**Gregorio Saldarriaga** is a faculty member in History and Coordinator of the Research Group in Social History the Universidad de Antioquia in Medellin, Colombia. His primary research interests center on the history and culture of food and drinking practices in the context of colonialism. He is the author of various articles on this subject. He has an M.A. and Ph.D. in History from the Centro de Estudios Historicos at El Colegio de Mexico.

**Gyorgy Scrinis** lectures in food politics in the Melbourne School of Land and Environment at the University of Melbourne, Australia. His research focuses on the sociology and politics of food production and consumption, with a focus on nutrition science, functional foods, biotechnology and nanotechnology. He is the author of *Nutritionism: The Science and Politics of Dietary Advice* (2013).

**Jerry Shannon** is a temporary assistant professor of Geography at the University of Georgia. He is broadly interested in how urban redevelopment centered on spaces of consumption affects the everyday lives of marginalized populations. His most recent project looks specifically at food deserts and how efforts to improve and measure food access interact with broader community development strategies.

**Heidi Zimmerman** is a PhD candidate and graduate instructor in Critical Media Studies in the Department of Communication Studies at the University of Minnesota, Twin Cities. Her work focuses on the way in which media— especially food and lifestyle media—participate in how citizenship, ethics, and environmentalism take shape within contemporary culture.

# Acknowledgements

The editors wish to acknowledge a number of people in the creation of this volume. First, we are incredibly appreciative of the Critical Food Studies editor, Mike Goodman, who encouraged us since the conception of this volume, was patient with various challenges that arose, and provided valuable feedback at multiple stages. We feel privileged to be a part of the series. Along similar lines, we are grateful to our contributors, many of whom played a central role in envisioning the volume and helping us to develop the tab system that we have used to organize the chapters. Our contributors also shared our enthusiasm for 'doing nutrition differently' and opened our eyes to seeing nutrition from multiple perspectives. We thank them deeply for their willingness to participate in this work, and for their efforts in shaping our vision for this collection.

We are also especially grateful for the feedback and suggestions that we received from several reviewers, including those that reviewed the volume in its entirety, as well as those that reviewed and provided comments on individual chapters. Editing and reviewing is a critical part of the work that goes into an edited collection, and we are indebted to those that took the time to help us with this process.

Finally, we are grateful for the feedback and support that we received from a number of our family members, friends, colleagues, and students. The momentum for this volume really derives from a series of informal conversations and debates about food, health, and nutrition that we have had with numerous others throughout the years, from which we came to recognize the need to envision how we might begin to 'do nutrition differently.' We are lucky and thrilled to be surrounded by a supportive and vibrant community that is willing and excited to engage in such discussions.

*To our parents:*
*Linda Hayes & Rusty Conroy*

# Introduction

Allison and Jessica Hayes-Conroy

In this volume we have a collective interest in nutrition – that is, the nourishing of bodies, the provision of food to living organisms (oneself or others). We understand this food-body relationship as something expansive, intricate, and diverse. It is certainly not something that begins and ends with "nutritional guidelines," although a quick Internet search for "nutrition" reveals the extent to which the term has become tied to "facts" and "guides." Hoping to move beyond facts and guides, this book seeks to offer multiple counter-discourses to what many critically-minded food scholars, feminist scholars and nutrition professionals have come to see as a narrow and repressive approach to diet and nutrition. In this introduction, we refer to this repressive approach as 'hegemonic nutrition,' while recognizing that it is more variable and less monolithic than such a designation seems to encourage. Yet, while hegemonic nutrition is variable, or perhaps we should say flexible (ever accommodating to ceaseless change), it is characterized by a few constant attributes:

First, while specific facts and guides may come and go, hegemonic nutrition rests on the assumption that food, and thus the food-body relationship can be *standardized*. As Nick Cullather's (2007) work on the calorie makes clear, the modern relationship with food has been charted by a dominant belief that "food" can be given "a uniform meaning . . . and a standard value that can be tabulated as easily as currency or petroleum" (339). Over the past century, our relationships to food (and to our food systems) have been quantified – e.g. calorie counts, nutrient counts, serving sizes and figures, Body Mass Indices (BMI), and so on – in an attempt to provide a universal metric for understanding the food-body relationship. While there have been a number of important reasons for such standardization, a few of which we recount below, it seems clear that the encouragement of more progressive and socially and ecologically attuned *nutritions* will require us to move beyond the presumption that our food-body relationship can be wholly (or even largely) understood through universal metrics.

Second, and building from our first point, hegemonic nutrition tends toward *reductionist* understandings of nourishment. The standardization of food's value has privileged the universal metrics – calories, nutrients, and so on – while redefining 'food' (at least for health purposes) to be the sum of these standard parts. Gyorgy Scrinis has consistently made this point (2002, 2008, and in this volume) in outlining what he calls "nutritionism" – an overwhelming emphasis on understanding the value of food in terms of nutrients, both macro-nutrients (like protein, fats, and so on) and micro-nutrients (like vitamins and minerals). Scrinis writes:

> Nutritionism is the dominant paradigm within nutrition science itself, and frames much professional and government-endorsed dietary advice. But over the past couple of decades nutritionism has [also] been co-opted by the food industry and has become a powerful means of marketing their products (Scrinis 2008, 39).

Scrinis explains that this focus on nutrients has become ubiquitous, and is implied, assumed, or taken for granted in much contemporary (mainstream) engagement with food, thus overwhelming other ways of encountering and experiencing food. Indeed, the reduction of food to nutrients comes at the expense of other ways of knowing the food-body relationship, perhaps particularly global South and indigenous valuations and assessments of food and health, which tend to be suppressed through the normalization of food as nutrients, and associated Western ways of knowing food (Waldron 2010, Galvez 2011, Also see Rodriguez, this volume). In short, as food becomes reduced to little more than measurable constituent parts, multiple layers of valuation beyond the nutrient are lost. As Scrinis (2008) writes, "The assumption is that a calorie is a calorie, a vitamin a vitamin, and a protein a protein, regardless of the particular food it comes packaged in" or, we would add, the social and ecological systems through which it is produced (41).

Our third point regarding hegemonic nutrition follows closely from the above: hegemonic nutrition is fundamentally *decontextualized*. Because it is based in standardizable and reductionist ways of knowing food, hegemonic nutrition is necessarily negligent about context. To be clear, this is not an argument about the placelessness of modern nutrition science (although it seems possible that such an argument might be made). Rather it is an argument, in the tradition of Donna Haraway (1988), about *situatedness*. Hegemonic nutrition pretends to know food from nowhere, while being applicable everywhere; its disembodied objectivity not only attempts to universalize the richness of regional cultures and the complexities of the human-food-environment relationship, but it also feigns sensitivity to place and the epistemologies of location by incorporating them into its calculable logic. For example, the notable attention currently given to "cultural appropriateness" among nutrition practitioners (see Frank this volume), at first seems to contradict hegemonic nutrition's decontextualized nature. Yet, when we take a look at what cultural appropriateness has come to mean, we often find superficial praxis – e.g., altering BMI recommendations for certain ethnic groups, or including a wider range of 'ethnic' foods in healthy eating guides. Culture is not central to the goals of hegemonic nutrition; it is an after fact, "stirred back in" to the reductionist food batter. So, while hegemonic nutrition may not portend one single prescribed way of eating, it certainly claims to know the central parameters of food-body relationship – that is, what is inside and what is outside of the scope of nutrition. This volume actively fights against such decontextualization. Collectively, our authors suggest that in order to understand the process by which a body is nourished, we need to understand the complex ways in which people, foods, lands and places come together. In this volume, it matters whether foods are derived from just social and environmental relationships or not; it matters what kinds of historical, cultural and emotional linkages foods have. Such considerations are not tangential approaches to nutrition, but rather, we argue that they are at the heart of the food-body relationship that nutrition seeks to study. In an attempt to move beyond hegemonic nutrition, our critical approach to nutrition does not bracket off environment, culture and heritage from nutrition science.

Our fourth and final point follows closely once again – due in part to its bracketing off of wider cultural, ecological, and social contexts, the knowledge system upon which hegemonic nutrition is based is deeply *hierarchical*. Hegemonic nutrition aggrandizes expert knowledge, which purports a denigration of other knowledges, unless they come to be known and accepted as legitimate by experts (e.g., fermentation has a long history of use in promoting beneficial aspects of the body-food relationship, but has only recently become accepted in mainstream nutrition as science and industry have together legitimatized the value of 'probiotics'). In claiming that hegemonic nutrition is linked to an epistemological hierarchy based upon 'expert' (often scientific) knowledge, however, we do not assert that hegemonic nutrition is produced and proliferated by mainstream nutrition science alone. There are a number of production and advancement points of hegemonic nutrition, which may include nutrition practitioners, the media, industry, and social institutions like school, family, and church (to name a few). Such sites become points of (re)production and advancement for hegemonic nutrition when they support standardization of the food-body relationship, maintain a reductionist understanding of food and health, decontextualize food behavior from its diverse and complex roots, and reinforce the hierarchy of expert knowledge.

We also want to be clear that while we recognize science – *nutrition science* – as a major player in the proliferation of hegemonic nutrition, we do not seek to attack nutrition science in this volume. Indeed, we do not see nutrition science as something to attack, but rather as an important yet partial knowledge/practice, and one that needs be in deeper dialogue with other, diverse food and health knowledges/practices. We noted previously that there may be progressive political reasons to want to standardize, decontextualize, reduce, and rely on scientific expertise in understanding food-body relationships. For instance, neighborhood groups have used the scientific demonization of high-counts of salt, fat and/or sugar in convenience store foods to fight for better food access in inner cities, and some native American communities have used the (pejoratively) high BMIs of the obesity "epidemic," as a rally point to encourage a re-emergence of fresh and traditional foods. Still, the pages that follow emphasize the need to move beyond standardization, reductionism, decontextualization and hierarchy in nutrition. They also, in doing so, tend towards a de-centered view of nutrition science – seeing science as vital yet incomplete.

Certainly, many nutrition scholars and practitioners already recognize limitations of nutritional science, and this recognition has led to more varied and variable practice, as a few of the chapters in this volume indicate. Nevertheless, too many of those entering the field of nutrition still come into it with rigid ideas of what and how to eat and moreover, with an implicit hierarchical power structure for how to promote healthier bodies (that is: listen to the nutritionist, and do what s/he says) (see Frank, this volume). Particularly troubling is the coupling of expert advice about nutrition with tacit criteria for determining individual fault in nutrition practice – usually some combination of lack of education, motivation, and unwillingness to comply with the 'rules' of nutrition. De-centering (modern Western) nutrition science from nutrition may help to dismantle this kind of power hierarchy. Thus one of the goals of this volume is to document some ways in which nutrition practitioners and activists are working to 'do' nutrition in less normative ways – in ways that seek to change the rules of the game and who is playing them. To be clear, the 'game' is not nutrition science itself but rather hegemonic nutrition at large – a structure and system of understanding the food-body relationship that is broader than nutrition science, but that is often given legitimacy through science. In critique, the volume both *questions* the role of science and industry at large in creating certain social imaginaries that preclude other ways of doing food and *imagines* ways of doing nutrition differently.

**Diverse Nutritions**

We have already stated that hegemonic nutrition may not be as monolithic and omnipresent as the above line of reasoning allows. Indeed, perhaps hegemonic nutrition only seems so vast because we continually (re)write its authority through our everyday lay, professional, and scholarly performances. In this way, hegemonic nutrition seems akin to capitalism – a system of organization and understanding whose assumed dominance veils multiple already-existing or imaginable alternatives. Accordingly, this volume seeks to make more room for already-existing or imaginable alternatives to hegemonic nutrition. Borrowing from J.K. Gibson-Graham's notion of 'diverse economies' – an ontological approach to economy that re-considers the supposed dominance of capitalism and re-imagines the economy as something much larger than typically discussed – this volume seeks to encourage a recognition of 'diverse nutritions,' an approach to nutrition that softens the hegemony of 'hegemonic' nutrition and opens up imaginative space for nutritional alternatives. The analogy is perhaps best revealed through Gibson-Graham's well-known depiction of the economy as an iceberg:

> The iceberg is one way of illustrating that what is usually regarded as 'the economy' (i.e. wage labour, market exchange of commodities and capitalist enterprise) is but a small set of activities by which we produce, exchange and distribute values in our society. This image places the reputation of economics as a comprehensive and scientific body of knowledge under critical scrutiny for its a narrow focus and mystifying effects. The iceberg opens up a conversation

about the economy, honouring our common knowledge of the multifarious ways in which all of us are engaged in producing, transacting and distributing values in this hidden underwater field, as well as out in the air (communityeconomies. org "Key Ideas" 2012).

Hence, the iceberg becomes a metaphor for critical scrutiny, broadening conversation and honoring common knowledge. In applying this metaphor to nutrition we contend that what is usually regarded as "nutrition" (i.e., the quantifiable science of providing bodies with the right amount of nutrients) is just the tip of the iceberg in terms of the processes and practices involved with sustaining embodied life through food. Thus our nutrition(s) iceberg (Figure I.1), opens up a conversation about the diverse ways in which people come to know, appreciate and practice sustenance and health through food. It encourages us to recognize all of those aspects of bodily nourishment that tend to remain hidden (beneath the water in the iceberg figure). It also moves us beyond the supposed hegemony of standardizable, reductionist, hierarchical nutrition – and its 'stir back in' approach to context – and encourages the continued rise of projects of nutritional alternatives, autonomies and experimentations.

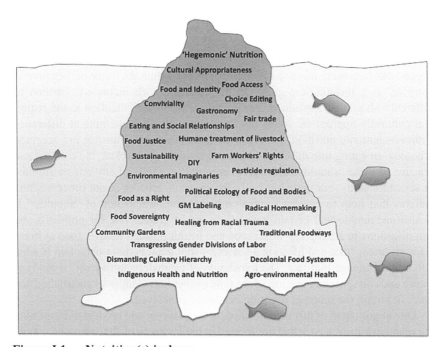

**Figure I.1    Nutrition(s) iceberg**

*Note*: Hegemonic Nutrition sits visible above the water while other aspects of nourishing bodies sit forgotten beneath the surface.

Part of the intent of a diverse nutritions approach would be (again parallel to that of diverse economies) to recognize the performative effect of the way we represent bodily sustenance through hegemonic nutrition; that is, by talking about and representing sustenance and health through food as largely a matter of standards and nutrient counts, it becomes harder to think, feel and value food in ways outside of this standardized food-body relationship. How and why is food good for us (if we don't have nutrients to explain it)? When and how should I eat (if we don't have diet guides to tell us)? Who is nourished and who is not (if we don't use scales and BMIs as indicators)? Without broadening the scope of what we understand as relevant to nutrition we are left with few answers to these kinds of questions. Thus, the project of diverse nutritions becomes one about altering performance: how might academics, professionals and activists *begin to perform new nutritional worlds, starting with an ontology of nutritional difference* (adapted from Gibson-Graham 2008). Moreover, what are everyday people doing already to creatively re-think and re-enact nutrition? And specifically, how might we recognize and sanction what is happening to make nutrition more just, more ecologically harmonious and more critically attentive to the politics of food, bodies and lands?

Along these lines, we recognize that the existing and increasing pliability of the 'rules' and 'guides' of hegemonic nutrition is a step in the right direction. Indeed, while we characterized the aforementioned attempts at cultural appropriateness as perhaps "superficial" – e.g., altering BMI recommendations for certain ethnic groups, or including a wider diversity of foods in healthy eating guides – this is not to suggest that becoming more sensitive to difference within the frame of 'hegemonic nutrition' is a meaningless goal. On the contrary, such increased attention to difference should be applauded. Yet we also want to call attention to the reality that culturally appropriate foods (and a number of other attempts at difference within hegemonic nutrition) generally do little more than translate pre-accepted "shoulds" of eating into different culinary languages. To be clear, the creation and practice of these "shoulds" themselves are not the "fault" of any one nutritionist or scientific study, but more generally the result of broader social processes and patterns that help to construct and to advance the anticipation of "shoulds." It is also undeniable that certain populations are at greater risk for problems and complications in attempting sustenance and health through today's food systems – for instance, "diet and lifestyle diseases," – but what is more uncertain is what will help to change that fact. The translation of ostensibly "universal" shoulds of eating ends up ignoring, erasing or confining other knowledges of sustenance that don't fit within the scope of nutrition, narrowly conceived.

One major intent of this volume, then, is to go deeper and broader in examining the relationships of power that are formed in the interactions that surround eating, diet-ing, and the sharing/provision of food, including the production of knowledge itself but also the material relationships (involving trust, judgment, shame, fear, belonging, hunger, satiation and so forth). This is a tricky task, to be sure, since we don't wish to deny the value of nutrition science, but we also want to explore the legitimacy of alternative knowledges, partial knowledges, and body knowledge, in order to make room for other ways of doing/knowing the food-body relationship, and to open the scope of possibility for what nutrition is, and how nutrition is practiced. The volume collectively hopes to convey that the way nutrition is currently envisaged and practiced is too narrow – too focused on the individual, for example, and on the formation of distinct "shoulds", whether culturally appropriate or not. In conveying this, we do not, however, seek to convert the reader to a different "should" of nutrition (e.g., one based in alternative food, or veganism, or decolonial diets, all of which are mentioned in the volume). Rather, the hope is to inspire more honesty about nutrition – to be conscious of judgments and fears on all sides, and also to recognize that nutrition inevitably involves relationships built upon uneven power, and that there are ways of practicing nutrition (both in professional and lay terms) that can act and re-act consciously to these realities.

## Audience and Scope

We see this volume as relevant to a wide audience including nutrition professionals, scholars and students interested in diverse aspects of food and nutrition, food activists, and anyone interested in thinking, feeling and valuing food in broader terms. Echoing what was said above, there is no singular conception of nutrition that this volume seeks to transmit. In fact, the various contributors to the volume are not necessarily in agreement with each other about the scope and definition of nutrition; we are, however, all accepting of the possibility of multiple, intersecting, and diverse nutritions. In this sense, rather than a structured map of nutrition, the contributors collectively offer a series of tracings that depict various elements of the food-body relationship that each has found important or significant. Our hybrid and partial tracings, when put together, are full of contradictions. For instance, we find that as we try to express and explore the broader contexts of food-body relationships, we come across conflicts in language. For some, it seems quite natural and appropriate to use the language of "clients" in discussing individuals that are seeking to improve their food-body relating. For others in the volume, however, the very use of the categorization of clients may indicate part of the problem. That is, not only does the language of clients point to potentially problematic assumptions about who has and who needs nutrition knowledge, but it also may work to erase broader socio-political contexts. This matter of language is just one example of a number of contradictions (and stylistic differences) encountered through the volume. As editors, we do not see such differences as a weakness. Instead, we embrace contradiction as part of the process of enabling diverse nutritions to flourish.

Our embracing of contradiction, however, does not portend a lack of coherency and consistency. Indeed, the contributors to the volume unite on a few central points:

1. The contributors agree that dietary practices need to be understood within the context of uneven relations of power both locally and globally. In other words, we acknowledge the existence of broader, structural constraints on individual patterns of consumption.
2. The contributors are dedicated to analyses of nutrition and diet that *question* mainstream assumptions about what healthy eating is, what a healthy eater looks like, and what a healthy eater should value. In this sense, we are aware of the epistemological struggles embedded within daily dietary practice, and the ways in which certain knowledges and knowers have become hegemonic.
3. The contributors seek to acknowledge the agency of individual actors in reproducing and resisting dominant regimes of nourishment, and wish to encourage alternative practices of nutrition that are effective and empowering to individuals and communities. Accordingly, the volume pays attention to the material ways in which bodies, food, and ideas come together to create opportunities for both stasis and change.

The desired outcomes of the volume follow from these key points. We hope the book will be not only theoretically useful for envisioning diverse nutritions, but also practically valuable in the provision of starting points for doing nutrition differently. Nutrition "intervention," broadly understood, is any initiative that attempts to create change in dietary behavior for the purpose of 'health' promotion, again broadly conceived. Thus, one intended outcome of this collection is to provide the roots for new ideas and new models of nutrition intervention/participation that 'do' nutrition differently, and in so doing, also 'undo' some of the hegemonic ways in which nutrition has been heretofore conceived and practiced. At the very least, we hope the volume will inspire an exploration of new ways to define and do nutrition that may steer us towards healthier bodies, communities, and ecologies.

**Tabs and Themes: An anti-structure**

In keeping with the volume's radical intent, we have opted for an anti-structure for the book. With the exception of this introductory chapter, the book's chapters have been organized alphabetically, like a bibliography. In place of sections, which typically provide the organizing structure of edited volumes, we have opted for a "tabs" system, in which thematic threads have been identified and highlighted at the start of each chapter. There are a total of ten tabs running throughout the book, thus enabling multiple ways of reading the volume. To be clear, any one chapter might correspond to four or five of the thematic tabs, but typically not all of them. In each chapter we have also left a number of 'blank' tabs, encouraging readers to come up with other ways of interpreting the volume. Below, we explore each of the ten tabs separately, in order to provide a background to the themes each represents. At the beginning of each chapter, the reader will also find an editors' note, summarizing the chapter's thematic and broader contributions to doing nutrition differently.

*Access*

The access tab follows authors who are engaging with varying aspects of food access, including economic and geographic access as well as less recognized aspects like emotional, cultural, or visceral access. At least six of our authors engage with matters of food access. In **Alkon**'s chapter "A Conversation with Navina Khanna and Hank Herrera," food access is a central objective of both of these food justice activists, but not perhaps in the ways traditionally espoused in the large literature on food access and food deserts. The conversation with Khanna and Herrera makes clear that access to healthy, nutritious food is not a simple equation, and it certainly is not about others telling low-income communities what they need and what kinds of food they *should want* to access. Rather, food access is achieved through legitimizing the experiences and knowledges of those who are "trying to make it," as well as through creating jobs, through capacity building, through removing the barriers that prevent people from becoming knowledgeable players in food systems, and ultimately through empowering people to reclaim their own communities and systems of food procurement and sustenance.

In a similarly critical yet distinct way, **Shannon**'s "Should we Fix Food Deserts?: The Politics and Practice of Mapping Food Access" questions the rigidity with which the notion of food deserts – that is, areas where fresh, nutritious food is difficult to find – has been understood. Shannon explores how the interpretations and analyses involved in food desert research tend to detract from complex scrutiny and solution-finding that could take into account mobility, diverse market types, difference and social stratification. Shannon's radical contribution to 'doing nutrition differently' is to insist that we will not solve the dilemmas of access to nutritious foods if we only focus on the kinds of conspicuous, large-scale solutions (like bringing in supermarket chains) that current analyses favor.

Four other chapters integrate the notion of access into their discussions. **Zimmerman**'s "Mobilizing Caring Citizenship and Jamie Oliver's Food Revolution" can be read through the lens of access. Zimmerman's analysis of the popular television series *Jamie Oliver's Food Revolution* (JOFR), suggests that a real revolution in eating habits cannot be understood within the frame of "free choice," but rather must work to solve the broader structural and institutional barriers that prevent people from accessing nutritious food – economically, geographically, culturally and viscerally/sensationally (actually physically desiring certain foods). Likewise, in **Frank**'s "Counseling the Whole Person" we see a parallel revelation about the necessity of attending to multiple constraints on food access; Frank suggests both that food access is central to matters of dietary habit and nutritional "choice," and that awareness of socioeconomic constraints on food access can be both challenging and edifying for nutrition practitioners. **Cadieux's** chapter on food gardens, "Other Women's Gardens: Radical Homemaking and Public

Performance of the Politics of Feeding," meanwhile encourages us to understand and interpret gardens as significant places where people work to appreciate and reconfigure the eco-social relations of food access. Finally, in "Doing Veganism Differently: Racialized Trauma and the Personal Journey Towards Vegan Healing," **Harper** raises the point of food access in a critical light encouraging us not only to recognize the ways in which economic, geographic and epistemological access to food has been racialized, but also to become more attuned to the specificities of real people accessing food, and not to be quick to draw broad assumptions about populations based on large demographic or geographic categories.

*Body*

The body tab traces the themes of body size, body politics, bodily health, embodiment, and viscerality through the volume's diverse chapters. Five of the volume's authors engage with matters of the body. Both **Beauboeuf-Lafontant**'s chapter "Our Plates are Full: Black Women and the Weight of Being Strong," and **Zimmerman**'s chapter on "Mobilizing Caring Citizenship and Jamie Oliver's Food Revolution," directly confront the subject of body size, albeit from different angles. While Beauboeuf-Lafontant utilizes understandings of how gender is racialized to explore physical overweight in American Black women, Zimmerman employs critiques of neoliberalism to counter the belief that obesity is largely a matter of personal responsibility. Both Beauboeuf-Lafontant and Zimmerman contribute to growing and broader discussion on body size, overweight and obesity that seeks to question not only the ways in which discourse on body size has been framed but also the multiple assumptions integral to the defining and analysis of obesity/overweight as an 'epidemic.'

Also contributing to this discussion, **Scrinis**'s chapter, "The Nutricentric Consumer," challenges measurements of overweight and obesity such as the Body Mass Index (BMI) as part of a broader paradigm of 'biomarker reductionism' that condenses bodily nutrition and health to a sequence of quantifiable biomarkers. Scrinis goes on to examine contemporary changes to the body-food relationship more broadly, suggesting multiple erasures wrought by a reductive focus on nutrients. **Newcomer**'s "Nutrition is" echoes these concerns and brings them to new ground, reflecting on the quantification of the food-body relationship via 'calorie counting,' and demonstrating the profound interweaving of mental calculation, physical activity, hunger, fatigue, instability, so-called 'disordered' eating, and bodily trauma on multiple levels. Similarly, **Harper** explores her powerful experience of the interconnection of racial trauma with a diet high in processed foods and bodily ill-health. In the end, both Newcomer and Harper uncover the possibilities of healing the minded-body through personal and political struggle to yield new ways of thinking and feeling food.

## Colonial

The colonial tab encompasses chapters that engage with issues of colonization, colonality, and legacies of colonialism. The tab includes authors who utilize the language and theories of decolonization in arguing for a way to 'do nutrition differently.' At least seven of the contributing authors engage directly with the colonial theme. **Fazzino and Loring**'s chapter, "Nutritional and Cultural Transitions in Alaska Native Food Systems: Legacies of Colonialism, Contested Innovation, and Rural-Urban Linkages," invite sensitivity and specificity in the conceptualization of nutrition among indigenous communities. Fazzino and Loring examine the relationship of Alaskan and federal politics and policies in curbing access to the traditional foods of Alaska Native peoples, and they discuss a widespread (albeit uneven) nutritional transition among Alaska Native peoples, which stems from deep-seated patterns of colonial control. One of the take away points of the chapter with respect to this theme is that food and nutrition policy must not be uniform, but instead must be "indigenized" through variable strategies designed to enable the thriving of indigenous health and wellness initiatives.

**Rodriguez**'s chapter "Another Way of Doing Health: Lessons from the Zapatista Autonomous Communities in Chiapas, Mexico" and **Harper**'s chapter on veganism both explicitly draw upon the notion of decolonization in describing the kinds of changes to dietary patterns that they and other scholar-activists seek. Rodriguez interprets the activities of the Zapatistas in creating autonomous health and nutrition projects as decolonial actions – actions aimed at de-linking from the intersectional hierarchies created by the colonial system. Harper details her own personal journey as a black racialized female subject in the USA, expressing how she came to practice and encourage body/mind decolonization through a plant-based diet. Harper focuses on the ways in which the racist colonial project intersects with patterns and epistemologies of eating, and she discusses the formation of her *Sistah Vegan Project*, a book anthology and online community of African Diasporic females whose vegan practices are firmly rooted in a de-colonial body politics. **Hayes-Conroy and Hayes-Conroy**'s chapter also revolves around the theme of decolonization, using lessons learned from feminist postcolonial scholarship to imagine what a feminist approach to nutrition might look like, and feel like. The chapter begins with a critique of "core" nutrition values, which are rooted in white/Western culture, and goes on to describe how this core can begin to become decentered by attending to a number of considerations that are born through decolonial activism, as well as feminist postcolonial scholarship.

**Harris**'s chapter "Traditional Knowledge and the Other in Alternative Dietary Advice," also strongly connects with the colonial theme. Harris examines the rhetoric of the Weston A. Price Foundation (WAPF), which argues for the importance of traditional and indigenous diets and serves as a challenge to hegemonic nutrition, but in doing so reproduces problematic and oppressive colonial discourses. Specifically, Harris explores how WAPF represents the 'primitive' as morally virtuous and living close to nature, and sees primitive diets not only as a route toward healthier nutrition, but as a way to avoid the moral decline of western civilization. While ostensibly interpretable as a voice for indigenous wisdom, Harris reveals how such arguments are founded in dichotomous Western thought, which establishes suspect divisions between nature and culture, primitive and modern, tradition and science.

**Saldarriaga**'s chapter, "Food, Community and Power from a Historical Perspective: Keys to Understanding Death by 'Lethargy' in Santa Maria del Antigua del Darien," also examines matters of colonialism, this time from a historical perspective. With respect to this theme, one of the valuable contributions of this chapter is to demonstrate how food, and with it varying notions of nutrition, played a significant role in the unfolding of colonialism in the Americas. Specifically, Saldarriaga explores how bread – baked carbohydrates – was seen as the foundation of nutrition and civilization according to the dominant (Spanish) culture, and how this privileging of bread and those who ate it contributed to hierarchical categorizations of different classes of human beings (according to their food habits). He also demonstrates how, ultimately, this nutritional hierarchy was fateful in the case of Santa Maria del Antigua del Darien.

Finally, **Cadieux**'s chapter, suggests that food gardens are land uses that manifest colonial histories in many different layers: in the layouts, the foods, the justifications, the rules, the social relations, and the aspirations encoded therein. Cadieux also points out that food gardens can be sites of resistance to both the ecological and social dimensions of colonality. She explains how gardens provide a crucial site for reflexive discourse on colonial relationships, perhaps especially due to the co-existence of conservative and progressive motives for, and modes of, gardening, and the sympathetic space that tends to be afforded by shared interest and shared practices in gardening. She thus adds to the volume's insistence that food, sustenance and feeding must be examined with respect to the multiple colonial relationships in place along the food chain.

*Discourse*

Although arguably notions of discourse are present throughout all of the contributions in this volume, through the discourse tab, we draw specific attention to authors that seek to highlight and emphasize the work of narrative and language in the production of power-laden knowledges of food and nutrition. There are nine chapters in the volume that weave in discussion of this broad theme in either direct or indirect ways. **Beauboeuf-Lafontant**'s chapter focuses on the discourse of 'strong' Black women, and the ways in which the performance of this racialized feminine construct takes a toll on the minded-bodies of Black women in the US. In speaking about the diverse ways in which Black women respond *with food* to the racial and gendered burdens placed upon them, Beauboeuf-Lafontant is seeking to amend the lack of attention paid to how ethnically and racially diverse women respond to particular cultural constructions of femininity.

**Harris**'s chapter also explores a racialized discourse – the challenge to hegemonic nutrition leveled by the Weston A. Price Foundation (WAPF), and particularly by their founding president, Sally **Fallon Morell** in "Ancient Dietary Wisdom for Tomorrow's Children" (this volume) – examining how it succeeds and falters in its resistance to hegemonic nutrition, and in doing so offering broader lessons for doing nutrition differently. One of Harris's most critical points is that while the WAPF elevates the importance of traditional and indigenous diets, it does so by basing its arguments in the discursive construction of the primitive other, thereby reinforcing racialized boundaries. Harris also examines the moral basis of WAPF representations of nutritional choices and constructively compares this discourse with that of another contemporary critic of mainstream nutritional science, Michael Pollan. One ultimate objective of the chapter is to open a discussion about how best to valorize traditional knowledge in a society that places great worth on ways of knowing based in Western science. Harris insists that if the West is to look towards other cultures for guidance about how to do nutrition differently, it is important to acknowledge the violence with which "other" knowledges have been appropriated in the past, and work to ensure such learning processes help to move us beyond colonial epistemologies and power structures.

We have already mentioned that **Saldarriaga**'s chapter is based in part in an examination of historical colonial narratives about what does and does not constitute proper nutrition (e.g. bread as the nutritional foundation for colonists, and a site for differentiating between colonists and "other" groups for whom baked carbohydrates is not a staple). Saldarriaga's chapter also explores how colonial discourses on food and nutrition intersected with and helped to bring meaning to notions of community and urban space. In these and other ways Saldarriaga's chapter is expressly about discourse – about the ways in which edible products are given meaning, about the ways in which the provision of food comes to represent political presence and power, and about the ways in which nutritional discourses have tremendous command over bodily health and well-being.

**Cadieux**'s chapter on food gardens is also uniquely about discourse; the chapter demonstrates how gardens are important not only as a site and topic of food and nutrition discourse but also as a *mode* of discourse. Cadieux argues that gardening is constituted by a series of practices that are both material and discursive, and the material and discursive aspects of gardening interact in ways that allow people to explore their interactions with components of the food system. Cadieux also insists that apt navigation of *gardening as discourse* can be empowering in ways that it is valuable to recognize – since that power can be mobilized progressively or exploitatively.

Finally, the chapters by **Shannon, Zimmerman, Harper, and Hayes-Conroy and Hayes-Conroy** all speak in different ways to the theme of discourse, and may be read as critical analyses of and counter-discourses to some common modes of knowing within hegemonic nutrition. Shannon counters the ways in which food desert research "fix" food deserts as objects of study, erasing significant factors involving urban space, mobility, and food market diversity. Zimmerman, meanwhile, counters the mainstream discourse on obesity, which argues that obesity is a result of personal responsibility and individual free choice, and she reveals how this standard discourse is steeped in neoliberal ideology. Harper counters the normalized whiteness and more overt racisms and racist erasures that have coincided with discourse on plant-based diets. Finally, Hayes-Conroy and Hayes-Conroy emphasize the importance of origin stories in decentering nutrition's white/Western "core," and look to narrative and storytelling as a way to practice doing nutrition differently.

*Emotion*

Related to matters of the body, but deserving a grouping of its own is the emotion tab. This tab highlights diverse connections of emotion with food and eating. At least five of the volume's chapters engage with matters of emotion. **Beauboeuf-Lafontant**'s chapter details the ways in which overeating becomes the outward expression of emotional states that have no direct mode of expression. She is directly engaged with matters of 'emotional eating' as well as the overall emotional well-being of Black women. Her discussion of suppression of emotion with food is echoed in **Harper**, who talks about trauma in ways that move beyond medicalization or pathology, and goes on to also suggest that physical and emotional healing may be possible through decolonial plant based diets. **Hayes-Conroy and Hayes-Conroy** also take up the topic of emotion through a discussion of decolonizing desire, attending to the ways that the material body comes to take comfort in certain foods but not others. The discussion particularly emphasizes the partiality and incompleteness of decolonizing desire, and cautions against projects of bodily authenticity and purity.

Like Harper, **Frank** also takes a more holistic approach to the food-body relationship, and thus a more holistic model to nutrition counseling. Hers is a model that takes into account a person's aesthetic, emotional, cultural, socioeconomic, religious/spiritual, and lifestyle influences on food habits. Frank also directly discusses matters of emotional or psychological eating, and like Beauboeuf-Lafontant and Harper, recognizes that for some, food has become an emotional battleground. Such a recognition is again echoed in **Newcomer**, who speaks about the emotional upheaval associated with experiences of 'disordered' eating, trauma, insecurity and ultimately, health and healing.

*Nature*

In general terms, the nature tab refers to a thematic thread that is organized around concern for the natural environment. This theme is present in at least six of our chapters, in each taking a slightly different form. The chapter by **Rodriguez** takes on matters of control over lands and environmental resources. Rodriguez details the physical and intellectual labor of the Zapatista communities in advancing autonomous health and nutrition systems that work beyond the grips of transnational life science and pharmaceutical corporations. Rodriguez's contribution is to demonstrate the complex interweaving of the Zapatista struggle over their ancestral territories with their commitment to decolonized systems of food, health and nutrition, including the knowledge base from which each is derived. The chapter explores how the defense and collective working of the land of the Zapatista's ancestors – cultivating corn, beans, *chiles*, coffee and traditional medicinal plants – amounts to non-violent resistance against the 'Empire of Money.' Also speaking about colonization, but from a different perspective, **Fazzino and Loring** explore structural and societal impediments to achieving healthy diets in their chapter on Alaska Native food systems. With respect to the theme of nature, their contribution is to express how dietary health has much to do with the ways in which environmental 'resources' are managed.[1] This chapter specifically demonstrates the impacts of fisheries management under the Yukon River Salmon Agreement on the foodways and dietary practices of indigenous peoples of Alaska.

The theme of nature also emerges in less defined ways in three other chapters. In the chapter by **Alkon**, the environment emerges through concerns over who has access to land resources for production, and who can access food that has been produced in ways that are socially just and ecologically appropriate. One strong undertone through the chapter is the importance of enabling communities to reclaim their ownership of food production, and thus their primary relationship to the planet. Similarly **Harper**'s chapter suggests the need for a veganism that can simultaneously resist racism, environmental degradation, and ill-health in the Black community. Yet, at the same time, Harper's chapter underscores how the white, middle-class rhetoric of ethical, green eating that she encountered during her college years worked to silence discussion on race and racism and turn her off to other ways of engaging with plant-based diets. Differently, in **Saldarriaga**'s historical chapter, nature is not yet seen as something to be 'saved,' but rather as something to be interpreted. He traces the consequences of colonizers' inflexible understandings of nutrition, which prevent them from recognizing the bounty of South American indigenous agroecology as edible. Ultimately, like the others mentioned above, Saldarriaga demonstrates the incredible power that food holds as the primary relationship between human bodies and the planet that sustains us.

Interestingly, **Cadieux**'s chapter on food gardens also engages with nature as something to be interpreted. In the chapter, the more-than-human assemblages of food gardens are shown to play a major role in the generation of food and gardening discourse. Through the chapter, we find that this may be especially true in public garden project contexts, where there is a collective witnessing and experience of the environmental challenges that impinge on people's intentions

---

1   We use the term environmental 'resource' here because the view of nature as 'resource' is the most common in policy circles, and tends to define the way that government and other institutions manage and relate to their land and water ecosystems.

for gardening and feeding. Such collective perception has varied implications for gardening, feeding and nutrition practice, as well as for eco-social learning and resilience.

*Race*

The race tab puts focus on matters of race and racialization across the volume, especially noticing the intersection of racial trauma with the food-body relationship. There are at least six chapters in the volume that attend to this important theme. In **Alkon**'s chapter, race is revealed as a central analytical point of the food justice movement – a movements that critically acknowledges the influence of race and class on the production, distribution and consumption of food. Alkon's conversation with two food justice activists illustrates some ways in which low-income communities and communities of color seek to create local food systems and ways of knowing food and nutrition that better meet their needs.

The theme of race also figures prominently in the chapters by **Beauboeuf-Lafontant** and **Harper**. As previously expressed, Beauboeuf-Lafontant's chapter is organized around the analysis of racialized gender roles – specifically those of US American Black women. Harper also analyzes through the lens of race, exposing not only her own experiences with racial trauma and the unspoken whiteness of plant-based dietary discourse, but also the ways in which racism and normative whiteness factor into her and others' projects of dietary decolonization.

Finally, two other chapters advance attention to race through their engagement with theories of coloniality and decolonialization. **Harris**'s analysis uncovers the unintentional reproduction of oppressive discourses of race and coloniality in the work of the Weston A. Price foundation (see **Fallon Morell**, this volume). Specifically, Harris examines the racial "othering" that happens through a white American/European appeal to "primitive" diets, and he argues that the appeal to the diet of a primitive other echoes feminist theorist bell hooks' notion of "eating the Other" – the manner in which constructions of the "other" are incorporated into Western subjectivities through consumption (hooks 1992: 21). Echoing Harris's call for decolonizing ways of knowing food, **Rodriguez**'s insistence on the decolonial imperative implies a obligation for anti-racist nutrition policy and politics, and specifically, a move away from racial hierarchies central to colonialism and the continued militarization of indigenous lands and peoples. Likewise, **Hayes-Conroy and Hayes-Conroy**'s discussion of decolonization and decentering also revolves around notions of racial hierarchy by examining the ways in which white/Western nutritional values are reproduced as universal and apolitical core truths. The chapter details the ways in which decolonization becomes an attempt to center whiteness as the supposed nutritional core.

*Science*

The science tab groups all those who reference matters of science and scientific knowledge production, including especially critiques of Western science, both explicit and implicit. At least six chapters in the volume speak to the broad theme of science. In her chapter, **Frank** solidly suggests that nutrition science is just the beginning of the practice of nutrition, and that nutrition counseling must be seen as a wide-ranging pursuit. Frank offers a holistic model of counseling that mediates science with a wide variety of aspects of socially-embedded lived experience. In a comparable, but more fervent way, **Scrinis** offers an expansive critique of the nutrient-centered practice of nutrition. While, as Scrinis imparts, nutrition science has up until now largely escaped critical scrutiny, the chapter seeks to redress this lack of inquiry. Instead of focusing on specific nutritional hypotheses and advice, however, Scrinis concentrates the taken-for-granted reductive focus on nutrients within nutrition science. The chapter outlines some of the characteristics of what Scrinis calls *nutritionism*, and examines how nutritionism has represented and shaped the minded-bodies of contemporary consumers.

Related to Scrinis' analysis of nutritionism, science comes into **Harris**'s chapter in the comparison he offers between two different critics of hegemonic nutrition – Michael Pollan and Sally Fallon (Morell). Harris points out that **Fallon Morell**'s counterargument against the dictates of dominant nutritional science rests on a divergent suggestion of what human bodies need – the argument that the body's need for animal fats is natural. Interestingly, and as seen in Fallon's chapter, the main way that Fallon backs these claims is by reliance on nutrient-centered science (studies that, she says, have been shirked by mainstream nutrition). Pollan, on the other hand, uses different rhetoric in his rejection of mainstream nutrition science, tending not to focus on nutrients, but rather on whole foods and cultural traditions. Through this discussion, both Fallon Morell and Harris contribute to debates over whether and how traditional knowledge should interact with hegemonic nutritional science.

In a less direct, yet still substantial way, **Rodriguez**'s chapter also contributes to the debate over the interaction of traditional and indigenous ways of knowing with mainstream Western science of health and nutrition. Rodriguez honors the work of the Zapatista communities in Mexico, whose autonomous projects attempt a decolonization of health and nutrition. He explains that Zapatistas always start with natural medicine/methods of healing but, if necessary, will also use chemical medicines and methods of Western medicine to manage an illness or trauma. Indeed, while struggling for autonomy, the Zapatista promoters of health and nutrition do not oppose modern health sciences *per se*, but rather they oppose the capitalist nature of modern health, medicine, and nutrition. Similarly, **Hayes-Conroy and Hayes-Conroy**'s chapter on feminist nutrition examines and critiques the role of science in the construction and legitimization of white/ Western core values, while also discussing the continued importance of science as a situated-yet-significant contributor to nutrition-based inquiry. The chapter concludes with a nod to science and its potentially radical functions. Finally, **Shannon** offers a critical look not at nutrition science, but at geographic and social science. The chapter analyzes the role that scientific research on food deserts plays in determining the course of proposed solutions to problems of access to healthful foods. Shannon's contribution is to encourage more critical and nuanced science, attentive to specificity and difference.

*Structure*

The structure tab primarily follows structural constraints both on nutrition (bodily sustenance through food) and on the possibilities of doing nutrition differently. The tab includes matters of social structure at large, especially patterns of social and economic stratification, as well as institutionalized norms, customs and ideologies. To us, structure most broadly refers to the persistent ways in which society is arranged, which shape and restrict the available options and opportunities for eating and relating to food. There are at least six chapters that speak to matters of structure in the volume. **Alkon**'s chapter points out from the start that the unevenness with which fresh healthy foods can be found/accessed across the landscape is produced by an array of political and economic factors. Alkon is particularly concerned with the institutional racism that has influenced US agricultural policy and land use planning, and has structured who has access to land, capital and government support, and who does not. Food justice activism (the focus of the chapter) is expressly interested in addressing such inequities and working with the communities that have been marginalized by these structural imbalances.

In similar terms, **Fazzino and Loring**'s chapter also speaks to the need to examine patterns of social and economic inequity and at the policies that create or reinforce this inequity in order to understand what true food security would look like with respect to Native Alaskan and other indigenous peoples. Meanwhile, the chapter by **Frank** references awareness of structural constraints as a vital element of her holistic model of counseling – arguing that sensitivity to socioeconomic constraints is a necessary precursor to providing nutrition counseling that is actually usable to an individual. **Cadieux**'s chapter on food gardens is also attentive to structure insisting on the value of gardens as places where people can explore and reorganize the ways in which our systems of sustenance are structured.

Finally, both of the chapters by **Rodriguez** and **Zimmerman** express concerns about the dominant economic structure and its impacts on nutrition and the capacity of people to do nutrition differently. Rodriguez explores the sentiments of the Zapatista communities that worry about collusions between 'the Bad Government' and the 'Empire of Money,' which appropriate the fertile lands of indigenous peoples and work against their autonomy in health and nutrition. Zimmerman's chapter examines the ideological consequences neoliberalism on the 'problem' of obesity and simultaneously, through this exploration, she exposes neoliberalism as the pervading economic structure, which prevents necessary change in the current corporate industrial food system.

*Women*

The women tab tracks a variety of themes directly relating to women including women's health, women's sexuality, and the construction of femininity. There are at least seven chapters that engage, directly or indirectly, with these themes. As has been referenced multiple times already, **Beauboeuf-Lafontant**'s chapter is organized around the analysis of racialized gender, and specifically around the discourse of the 'strong' Black woman. Her contributions to women's and gender studies come in her intricate study of the ways in which a particular racialized feminine construction impacts the food-body relationships of women. **Harper**'s chapter shares many similarities with Beauboeuf-Lafontant, as she details her own scholar-activist journey to decolonized plant-based diets. The intersectional oppressions against which Harper struggles are shown to be particularly detrimental to the health and well-being of women's bodies, and thus the political, emotional and physical struggle for a vegan diet, for Harper, becomes all the more vital. Echoing Harper's discussion of trauma and healing, **Newcomer** also focuses on the ways in which nutrition intersect with the emotional and physical well-being of women. Newcomer's words highlight the kind of complex and situated struggle that many women experience with respect to their food-body relationships.

**Cadieux**'s chapter focuses on the central role that women play in food activism and particularly in food gardening. Through the chapter, Cadieux is interested in exploring how women navigate the tension between collective or state-centered approaches to food system governance and more individual and/or neoliberal approaches based in personal responsibility. She is particularly eager about feminist approaches to understanding women's gardening/feeding experiences, which she says encourage more reflexivity regarding a number of key tensions found in practices of sustenance. The tensions she highlights include political withdrawal versus political engagement, and complicity in exploitation versus radical confrontation of food system hierarchies. In this way, Cadieux's case studies speak to a reflexive engagement with conflicts that have been highlighted in the critical literature on women's food and nutrition practices.

**Hayes-Conroy and Hayes-Conroy**'s chapter also contributes strongly to the theme of women in at least two ways. First, the chapter underscores how the lessons learned from feminist activism in the US and abroad can help to steer us towards more inclusive perceptions and practices of nutrition. Second, the chapter also specifically addresses the ways that recent scholarship on women and food – from issues of production to distribution to consumption – can help us to admit the impossibility of unified, coherent answers to the question "what's good to eat?," especially when we examine the intersections of gender with race, class, sexuality, age, ability, and other forms of social difference and inequity.

Finally, note that in both the chapters by **Fallon Morell, Rodriguez**, and **Zimmerman** women are shown to play principal roles in the food systems that each describes, although the connections to women's studies remains implicit. Fallon Morell's chapter reveals the discourse through which she has become a strong voice in the role of nutrition researcher, homemaker and activist, inspiring other women to follow in her tradition. In Rodriguez's writing, the Zapatista *mujeres* are described as central to the success of their autonomous projects of health and nutrition and to their struggles over land and ancestral territory more broadly. In Zimmerman, the interface between the women working in school kitchens (dubbed 'lunch ladies') and Jamie Oliver in his own television series, invites further scrutiny.

## References

Community Economies (2012) "Key Ideas" Iceberg image by Ken Byrne, accessed at http://www.communityeconomies.org/Home/Key-Ideas May 10, 2012.

Cullather, Nick. (2007) The foreign policy of the calorie. *The American Historical Review*, 112 (2).

Galvez, Alyshia. (2011) *Patient Citizens, Immigrant Mothers: Mexican Women, Public Prenatal Care, and the Birth Weight Paradox*. Rutgers University Press: New Brunswick, NJ.

Gibson-Graham, J.K. (2008) Diverse Economies: Performative Practices for 'Other Worlds'. *Progress in Human Geography*, 32 (5), pp. 613-632.

Haraway, Donna. 1988. Situated Knowledges: The Science Question in Feminism and the Privilege of Partial Perspective. *Feminist Studies*, 14 (3), pp. 575-599.

Scrinis, G., "On the Ideology of Nutritionism," *Gastronomica*, 8 (1) February, 2008, pp. 39-48.

Scrinis, G., "Sorry Marge," *Meanjin*, 61 (4), 2002, pp. 108-116.

Waldron, I. (2010). The marginalization of African indigenous healing traditions within Western medicine: Reconciling ideological tensions and contradictions along the epistemological terrain. *Women's Health and Urban Life*, 9 (1), pp. 50-71.

# Thematic Tabs for Chapter 1

- *Access*
- *Environment*
- *Race*
- *Structure*

Sample Tabs for Chapter 1

Chapter 1

# Food Justice and Nutrition:
# A Conversation with Navina Khanna and
# Hank Herrera

Alison Hope Alkon

**Editors' Note**: This chapter brings together the themes of *access, environment, race*, and *structure* in exploring the topics of health and nutrition with two food justice activists – Navina Khanna and Hank Herrera. The chapter's contribution to doing nutrition differently is to express why it is crucial to bring matters of institutional racism and economic inequity into any discussion of nutrition. Alkon's exchange with Khanna and Herrera also makes clear that we must attend to the divisions that are produced through mainstream nutritional rhetoric that sets up some as experts while low-income and minority communities figure as lacking in knowledge about their own nutrition.

The past 50 years have seen a dramatic increase in demand for locally produced organic food. In its earliest stages, this food movement was largely ecological in nature and accompanied the rise of the environmental movement (Belasco 1993). It was also part of a broader countercultural envisioning of alternative lifestyles. The first food to be explicitly called organic was grown by those who had gone back-to-the-land to create a more sustainable and interconnected community (Guthman 2004). In the cities, food cooperatives, farmers markets and community supported agriculture arrangements emerged to distribute this food.

By the early 2000s, the movement was no longer fringe. Organic food sales skyrocketed from a mere $1 billion in 1990 to an estimated $20 billion in 2007, representing approximately 20% growth per year (Organic Trade Association 2007). Farmers' markets became the fastest growing segment of the food economy, Nationwide, there were 3,596 farmers markets in 2004. By 2006, there were 4,385, representing an 18% increase (USDA 2006).

And yet, there are many neighborhoods—particularly those inhabited by low-income people and people of color—where fresh produce is wholly unavailable. This unavailability is the result of a complex array of political and economic factors. Institutional racism colors the history of US agricultural policy and land use planning, creating differential access to land, capital and government support (Alkon and Agyeman 2011). These policies determined who could farm versus who could only work as agricultural laborers, and the conditions under which that labor would be performed (Minkoff-Zern et al. 2011). Policies also determined who could buy a house in which part of town, which areas would receive adequate city services, and which would be zoned for toxic land uses or destroyed through eminent domain (McClintock 2011). As a result of such policies, affluent people and whites have largely had access to healthier environments and opportunities for ownership while low-income and people of color have not.

Against these bleak circumstances, a new strand of the food movement has emerged around the concept of food justice. Food justice activism consists of communities that have been marginalized by the above policies exercising their rights to grow, sell and eat food that is fresh, nutritious, affordable, culturally-appropriate and grown locally with care for the well-being of the land, workers and animals (justfood.org, White 2010). While other food movements have been criticized as color-blind or implicitly white (Guthman 2008, Alkon and McCullen 2010, Harper 2011), the food justice movement contains an analysis that recognizes and problematizes the influence of race and class on the production, distribution and consumption of food. Communities of color have time and again been denied access to the means of food production, and, due to both price and store location, often cannot access nutritious food. Through food justice activism, low-income communities and communities of color seek to create local food systems that meet their own food needs.

Given that low-income communities and communities of color experience disproportionately high rates of diet-related diseases, and are often the target populations for nutrition interventions, it seems that any attempt toward *Doing Nutrition Differently* should include the perspectives of food justice activists who hail from and work with these communities. What follows is a guided conversation between two such activists about their work and their approach toward health and nutrition. As a researcher studying food justice, I selected participants from among those I have worked with. I distributed questions in advance via email, asking for feedback. This allowed the activists an opportunity to shape the direction our discussion would take.

*Navina Khanna* is a co-founder and the Field Director of Live Real, a national initiative dedicated to amplifying the power of young people shaping a radically different food system, and a Movement Strategy Center Associate, where she works to build a more aligned and strategic food justice movement.

Her commitment to creating equitable, ecological food systems runs deep: Navina has spent over twelve years working to transform local, regional, and national agri-food systems from field to vacant lot to table—as an educator, community organizer, and policy advocate.

Navina holds an MS in International Agricultural Development from UC Davis, where she developed curriculum for the first undergraduate major in sustainable agri-food systems at a Land-Grant University, and a BA from Hampshire College, where she focused on using music and dance for ecological justice. She is also a certified Vinyasa yoga teacher and permaculturalist. A first generation South Asian American living in Oakland, CA, Navina's worldview is shaped by growing up—and growing food—in the U.S. and in India.

*Hank Herrera* is general manager of Dig Deep Farms and Produce. Hank has worked on issues of food systems and economic development for a wide variety of organizations in New York and California. His work has included the design and implementation of food systems in low-income communities, participatory action and evaluation research, capacity building and training. He is an MD, a psychiatrist, and has been a Robert Wood Johnson Clinical Scholar and a Kellogg National Fellow.

**Alison**: Tell me about the work you are currently doing.

**Hank**: I work for a project called Dig Deep Farms and Produce, which is a project of a not-for-profit organization called the Alameda County Deputy Sheriff's Activities League, or DSAL. DSAL was founded by a sergeant in the sheriff's office, Sgt. Marty Neideffer. [It is located in] Ashland, a part of unincorporated Alameda County. [Alameda County also includes the San Francisco Bay Area cities of Oakland and Berkeley].

A little over a year ago, some people in the community approached DSAL to be fiscal sponsor for some kind of gardening and greening project. One of these people gave Marty a copy of Van Jones' book about the green economy and the light bulb went off in his head. The book talks about creating living wage jobs for people that don't otherwise have them. The mission of DSAL is community betterment and crime prevention. Job creation is a major strategy for crime prevention.

I got involved with this gardening group and it all came together. Marty and his project director, Hilary Bass, raised $145,000 to hire community residents as urban farmers, for my salary, and for a farm manager salary. In April of 2010 we started Dig Deep Farms. We hired 10 people from the community, just everyday people. We trained them, planted out our first farm space and started our farm.

A large part of our funding was stimulus funds, which we expected to have for one year. However, two months after starting, we learned that there would be no more stimulus money, so we had to make a decision whether we were going to be a real business or not. Our goal is to create living wage jobs and all the people in the crew were investing in what they were learning to do. So we decided to start a CSA [community supported agriculture] to serve the people in the community, which is a low-income community of color.

[Our work is] about healthy food and it's about jobs and it's about people who grew up in the community—everyday people—having an opportunity to be part of this and eventually become owners.

**Navina**: I work with a couple of different organizations and related initiatives with the goal of building a stronger, more justice-based food movement. One organization is Movement Strategy Center, which builds alliances within different justice sectors. There, I'm doing a landscape assessment of the current food movement and using that to develop strategies for how we can create a winning movement.

**Alison**: What's a landscape assessment?

**Navina**: A landscape assessment is a socio-political mapping of who are the different groups that are working on food issues. What approaches do they take? If they're a member-based coalition, who is their membership? What's their sphere of influence? What kinds of strategies are they using and what's their approach? Are they about preserving farmers rights? Are they about food justice? Are they about food security? Are they anti-hunger? So, I'm trying to map some of that out. There are hundreds of community-based organizations that are doing food work and so we're looking at where are all these people and how can we bring people together to work towards collective action. So with Movement Strategy Center, the hope is to work with others to develop a winning strategy for our movement and build stronger alliance between groups.

The other project I work on is called Live Real, which is fiscally sponsored by the Food Project. The goal of Live Real is to build the capacity of low-income youth of color engaged in food justice to lead the food justice movement. This includes developing leadership skills, building relationships and building capacity for collective action. Most of my work with Live Real is campaign development and organizing. We are specifically using the 2012 [U.S.] farm bill as an organizing target. Not as the end of it all, but as our first organizing target.

**Alison**: How do issues of health and nutrition fit in to both of your work?

**Hank**: I saw your question when you sent them to us. I almost called you and said Alie, we don't talk about health and nutrition!

It's almost peripheral. We don't focus on it. In fact, Katie Bradley, a graduate student from UC Davis working with us, once asked one of our people—we were driving somewhere and he was telling us all about what's going on in the community—the lack of access to healthy food and what's in the community in terms of fast food restaurants and obesity and health disparities. He was telling us from his own experience. Katie then asked, what do you think about food insecurity in the community? He responded, what are you talking about? The "movement" language had no meaning to him.

**Navina**: So glad you're saying this Hank!

**Hank**: People where we work are living everyday lives just trying to make it. And all this other bullshit [the specific health/nutrition jargon] just doesn't matter to them so much. If we ask them, do you like fresh food and do you like healthy food and yes, they do. People know that they should eat healthier. One of our guys smokes like a stack. He's sick now. He's out of work. He was in the hospital. People understand that at some basic level the relationship between healthy food, healthy living. But it's about jobs for them. It's about making the community better.

The people that we work with, they get it, but they don't know it in terms of health and nutrition. Lamont (at the time, a 19 year old, African American employee of Dig Deep) talks about how people in the community are obese or they have illnesses and they need to have healthy food and how there's no healthy food outlets in the community. There are no supermarkets. He can lay it all out. He can put it all out there. But it's not in the terms that the movement people would understand.

**Alison**: You said that Lamont talks about food and nutrition but not in the same terms that the food movement is using [including many motivated by food justice]. What are some of the terms that don't work for him and what are some of the terms that he uses instead? Or not instead, but what are some of the things that he wants to emphasize?

**Hank**: He talks about not knowing what some of the things we're growing are. He never pronounces bok choy right. He says some other crazy thing. But he talks about recognizing that that's food that people normally don't have. He talks about lack of healthy food and how people are exposed to the chicken place and McDonald's and the taquería… Food that's not healthy.

**Navina**: So what's the question again?

**Alison**: How do health and nutrition play into your work?

**Navina**: So I think I would answer that question a couple of different ways. To me, the thing that motivates me, the reason I started doing food and agriculture work, is because of our disconnect with life and our planet in general. Not being in sacred, respectful relationship: not being in right relationship with ourselves or the earth or each other. And I think that's a lot of what is hurting with all of us right now – we're missing that. We don't live in right relationship. And so the idea of health is much bigger and is manifesting as a lot of different symptoms: a lot of physical diseases that we're experiencing. It's manifesting in the air pollution and the oil spills and the fact that we sit behind computers all day. It's manifesting in a lot of different ways. And I'd say that ultimately, my hope, and the reason why I do this work is that we can reclaim that relationship. We can reclaim our lives.

Hank and I both use this idea that transforming our food system and working towards food justice is about reclaiming civic and economic ownership of the system. And to me, that's a primary goal. The reason why food became the way, for me, to reconnect us to right relationship is because it's our most tangible connection to the world. The fact that we, way back in the day, started using agriculture as our way to procure food was one of the turning points in the way that we related to the planet.

So the reason why organizing and civic engagement work is important to me is that there are so many barriers that make it impossible for people to grow their own food, to know where their food comes from, to know what policies are shaping whether or not they can access good food or whether they can grow food or whatever else. And so I feel like part of my responsibility is to work to demystify some of that, and to create avenues for more people to be engaged in that conversation.

In terms of where I focus, when I was talking about Live Real and low-income youth leading the movement, the statistics are crazy. The health statistics, based on lack of access to real food, are just appalling. People predict that half of the youth of color born this decade will have diabetes. That's crazy to me.

We know that no social movement in history has ever succeeded unless the people who are most hurt are leading that charge. It's crazy to think of what's being predicted, and it's so important that we use this opportunity in these next few years to change that. To make sure they know what's happening and how they're being oppressed by the current food system, and the systems that perpetuate our food system (it's not just the food system, right?). And that they have the tools and the capacity to change that.

So I guess the other way that nutrition plays into that ... Just thinking about how so much of our food is processed, packaged food. That people don't have access to real food. And only 2% of food that's sold in the US is actually fair and healthy and ecologically grown. So about who's leading change but also making it clear to all of us that none of us are going to be living well unless we're able to change that too.

**Hank**: Can I follow up on that?

**Alison**: Sure.

**Hank**: I just want to riff on what Navina said about what moves her. I do this work because, where I grew up in San Jose, I was a Mexican kid who didn't speak with an accent. And I went to an all-Mexican junior high school. And I got picked out to go to advanced classes for this, that and the other. I'm convinced that it has as much to do with the fact that I didn't have an accent as anything else. I know that there were kids in that school who were smarter than me but never had an opportunity. And to watch kids get wiped out is intolerable to me ...

Lamont is like one of my kids, like one of my own children. If you saw him, you might see his tattoos first which go all the way up his sleeves. But if you really saw him, you would see that the inside of his heart and his soul is this very sensitive little boy who never had a chance. And yet, in spite of all that, he does what he does. He goes and talks at Rotary Club, he works on the farm, he's learned how to farm. He is learning to do sales. He does work in the office and we're trying to get training for him. On his own, he and his buddy Roy decided to go to Chabot [community college]. We didn't tell them to go to Chabot. What we did tell them is that if you guys learn how to do this business then someday it will be your business.

It goes so much beyond food. It's the local economy. And our people are getting it. It's about them and their community and whether they can turn their community around. It's civic engagement and building capacity. It goes so far beyond the food that we're growing. And we grow beautiful vegetables. We grow some of the most beautiful food around.

**Alison**: Hank touched on this a little, but are there mainstream or food movement ideas about "proper nutrition" that get in the way of what you're trying to do?

**Navina**: There are a few things. The two that immediately come to mind are that nutrition, as a field and as a funded intervention, is all about changing individual behavior. It doesn't get into systemic issues at all. There's this whole idea that people need to have better priorities, they need to make better choices. But people can't make better choices the way that things are right now in terms of economic and power disparities. That just needs to change. There's also this concept in traditional nutrition that uses crazy alternatives, like low-fat alternatives that are still highly processed. We believe in real food—food that truly nourishes people and the planet.

But even within the food movement, there seems to be this idea that there is a certain kind of cuisine that everyone should be eating. I mean, yes, everyone *should* be able to access heirloom tomatoes and kale and collards if they want to, but the way that our families have eaten for hundreds of years is also really healthy. And that cultural history is rarely part of the conversation. [Sometimes] it's a nutrition standard like when they give us pre-packaged enriched, white flour spaghetti instead of whole foods. And then there's the idea of what whole foods we should be eating... salad is not what my people eat; we haven't because of water problems for hundreds or thousands of years. We don't eat salad or enriched pasta, but we eat good food. Our cultural foods need to be part of what we're promoting as healthy nutrition.

**Hank**: Indigenous diets

**Navina**: Uh-huh

**Hank**: Leave us alone.

**Navina**: Right. And speaking of indigenous diets, think about the WIC programs and the food stamp programs and the kinds of food that you can get with that. WIC especially. You can only get crap with WIC except for eight dollars. Eight dollars you can spend at the farmers market and the rest, only crap.

**Alison**: Wow

**Navina**: It's like, you get government cheese. You get Post cereal. You get crap. Why can't people spend that money on good food or why can't that money go towards being able to have a garden or get a CSA or whatever it is? Real food of some sort. Be able to use it at the grocery store instead of the WIC office.

**Hank**: One of the cool things we're doing is that [Alameda County department of] Public Health is funding doctors to prescribe a food box to pregnant women, and we're going to be the provider of the food box.

**Navina**: The other thing, and maybe this is getting back to the food movement but there's the individualistic bias where they think that everything is a choice. Then there are some people that realize that there are food vacuums where people don't have access to food, so they promote putting in grocery stores in peoples neighborhoods. As if that's going to solve the problem when real the problem is the economic disparities. Like Hank was saying, if people don't have jobs ... If those grocery stores don't come with jobs, that doesn't help anything. Ideally, the stores would be owned by people that live in the community.

Here is something I've wondered about a lot: Chinatown is full of wonderful abundant produce. It's mixed use development—it's everything you could possibly dream of in terms of good urban planning and good access to food, walkability, all of those different things. But if you look at Oakland, and I'm sure this is true of a lot of different places, some of the highest poverty levels are in Chinatown. The reality is that people who live there aren't making good wages, and they're sometimes taking care of big families. And so even though they have access to all of that good food, they can't afford it. Not that bok choy is on the food pyramid, anyway.

**Alison**: Hank, a minute ago you said something about indigenous diets and about leaving people alone. I want to push that a little more because given the health and greater economic and social conditions that exist, what would it look like if left alone and what kinds of changes do you think need to be made versus what needs to be left alone?

**Hank**: There's a community in Arizona, the Tohono O'odham community, which has the highest rates of diabetes and obesity in the world. Before 1960, type 2 diabetes was virtually unknown among Tohono O'odham people. The introduction of industrial food and the rise in diabetes happened at the same time. About 10 years ago, they started to recreate their own indigenous food system with dry land farming, tepary beans and other desert foods. Traditional foods protect against the onset of diabetes.

If you look at the graphs that show the increase in the rate of childhood obesity, the inflection up starts somewhere between the late 60s, early 70s. That's the time when the impact of the industrial food system started to become pretty intense. That was the time of introducing high fructose corn syrup. Probably a decline in the amount of cooking people did at home and an increase in fast food and convenience food diets.

It's like, I was at a meeting once in Rochester where some nutritionists from USC were rescuing vegetables for the food banks. I walked up as this one nutritionist was saying "I was talking to those people, meaning Mexican Americans, about how they make tortillas and I was trying to show them that they didn't need to use lard. I don't know why they use lard." And all I could say was "because it tastes better." And it turns out, there's nothing wrong with real lard, with animals that are properly raised.

**Navina**: And in the proportions that we have traditionally eaten those things. We never ate the crazy ways that we do now.

**Hank**: So the beans, squash and corn diet is a complete diet. Just leave us alone.

**Navina**: That's the thing too. The nutritionists, their solution was use Crisco, then use margarine. All this crap that's not real food at all. It's not only moving us away from our cultural traditions and food that's actually healthier for us, but the only way that this food exists is through the industrial food system. It's in bed with the corporations that control our food system.

I have access to so much nutrition information. I know that quinoa is really good for me. I know that all these different grains are good, that brown rice is something that I should be eating. But my family grew up on white rice and that's what I eat, pretty much every meal because that's just how we do it. But that cultural aspect of things is never part of the conversation.

**Alison**: I know you both work with youth and community members. Do you try to have conversations around how people eat? What are some of the ways of talking about these kinds of things that works to connect with people and what are some that don't?

**Hank**: I was just talking with Katie about this and she tries to talk to people about what they eat, but they don't want to talk to her. She's probably going to change her research to focus on jobs. I will say to Lamont or Roy, when they go to McDonald's, why are you guys eating that stuff, but that's what they eat.

I think that for me, it's not a matter of, ok, here we come, we're going to grow this healthy food and people are going to immediately go, oh, now I know what I'm going to eat. Change will happen, but it's not going to happen because we give a lecture or admonish people. It's going to happen because people are going to assume their own habits that are better for them from a health point of view then the habits that they have now. And I'm fine with that. This is not a six-month process. This is a long-term project and process.

**Navina**: Relatedly, in my experience, showing, not telling is what works. I used to run a produce stand at an elementary school in Oakland. And I'll never forget the look on kids faces when they taste their first peach. They never even knew what a real peach looked like and then they taste their first one. Same with strawberries and other fruits.

Even parents who would come and see the produce there: they'd eye it the first time, then they'd come up because they were curious about the stand. Then they'd ask a question about some vegetable they had never seen before and then the next week they would decide to try it.

It's also not about judging people or about telling them how to eat differently, it's based in relationships. The same way, [a friend] and I were in the car together and he was like, you know you follow really close behind other cars. We laughed about it because he's been wanting to say that to me for a really long time because it's a bad habit that I have.

It's the same with youth that I work with or with other people. After a while of hanging out, we trust each other. Then I can say, "do you know that there's a cup of sugar in that can of soda that you're drinking? Is that really what you want?" Ultimately it's about us being in relationship with each other. And cooking food for each other and also being ok with the fact that yeah, I eat pizza sometimes. I eat cookies too.

**Hank**: It's funny, my first year of medical school, one of the first lectures was from an internal medicine doctor who said, you know, people buy all these vitamin supplements, but if they just ate healthy food three meals a day, they wouldn't need any vitamin supplements because our food should carry all the vitamins we need.

I started gaining weight in medical school, which is the same time McDonald's opened up across the street. And I ate a couple of Big Macs. I went to my internship and had my physical exam and the doctor said, you know, you're a little overweight but you're going to lose weight doing your internship. I gained weight. That's when all the crap was in the vending machines and you ate out of the vending machines. All this central adiposity is all about hormones that are in chicken and hogs and beef. It's about antibiotics that are in those animals. It's about chemical fertilizers and pesticides. We all carry this huge toxic load in our bodies. It's horrifying to me that pregnant women and breast milk have stuff that's already going to poison our babies before they even get born. I think its genocide.

**Navina**: It is genocide.

**Hank**: What are we doing? Why do we let this happen? I don't know what to do. Will it do any good to write to elected officials to ask, how can you let USDA do this? We're killing each other.

**Hank**: That's part of our whole thing. It's not a public health approach, if you know what I mean. All the ways that public health does stuff. We can't use those messages that are the official prescribed messages from the official prescribed institutions. It has to come from what happens down on the ground.

**Alison**: For the kind of health and nutrition that you're describing to be possible or widespread, what else would have to change?

**Hank**: We have to start creating space in communities to produce, pack, package, process (in terms of canning and preservation, not process in terms of making new things) the food that people eat in that community. And use that cycle as an economic engine so that you not only get healthy food that's whole food, but that the money that's involved with producing and selling food stays right where people are spending it so that it can do other things in the community.

**Alison**: That sounds almost exactly like what I would expect somebody to say that works on food issues in a much more affluent community. The idea that we have to do everything locally, that the community has to be involved, the money has to stay in the community. That sounds less different from the food movement than I would have expected.

**Hank**: Well, the difference is that I'm not talking about affluent communities. I'm talking about low income communities and communities of color. I'm talking about excluded, oppressed communities where the means of production and exchange is always ripped out of peoples' hands. The justice part is restoring ownership and control over the means of production and exchange and not letting the control continue to reside in the hands of affluent people.

**Navina**: Just to go a little bit further with what Hank is saying and what you're saying is that yes, I think that many different people may have the same analysis. It's all about local economy, it's all about local food, it's all about preserving our traditions, even. For [affluent communities] it might be about preserving the family farmer. The problem is, so everybody's losing in our current food system, right? But some people are losing more than others. And in the kind of communities that Hank is talking about in particular, people don't have the kinds of institutional power that they need to be able to change those things.

So for example, if you are living in poverty, and you are a tenant or on section 8 housing, you don't have land that you can use to grow food. Even if you saw a plot of land that you want to use. We know people in West Oakland who tried to use land in their own community to grow food, but because of their skin color, and the fact that they're on a piece of vacant property, they're considered to be loitering when they're just trying to grow food. Whereas outsiders with skin privilege can come into West Oakland and try to do the same thing, and it's called "community revitalization."

The same is true if you want to open a store in your own community. If you don't have access to the capital to get a loan ... Even though there's all these great community food projects grants, who has access to that funding to make it happen? If you had a shitty education because of the education system, how are you going to be able to read those things?

How do you have the time to apply? They're crazy to navigate. I think that a large part of [movement building] is thinking across different sectors and making the kind of health that we're talking about a priority.

A lot of it is change that will only happen in peoples' hearts and minds, and a lot of that's only going to change once people have experience and exposure to the struggles that others are facing. Some of it happens on an individual level. It's individuals that uphold systems, and that can choose to change things.

But it also happens on a bigger scale. For example, if our planning department really worked in concert with our public health department, and the office of economic development … If all three of those departments really focused together on creating the kind of healthy communities and jobs and local ownership and community controlled food systems and ecologically integrated agriculture the way that we envision it, we could probably get some things done. But right now all of it is very piecemeal and [these things are] not peoples' priority. People are thinking very short term and thinking about appeasing their constituents.

If people aren't questioning the systemic power disparities, they often get stuck in a service mentality and the approach becomes "well we just need to get food to these people" without thinking about "well why is it that only a certain kind of people are in need right now?" and "what needs to actually change for a group of people to not be in need?"

**Hank**: I agree.

**Alison**: Is there anything else that you all want to add or feel like was missing?

**Hank**: Navina was talking at the very beginning about restoring our connection to the earth, to life. Paul Farmer in his book *The Pathologies of Power*, talks about structural violence and structural inequities. Those inequities are not randomly distributed in the world's population. They're selectively distributed. You can see that in Alameda County.

And it's very frustrating to me to have all of our friends, [who are] good people, focused on changing policy and working at high levels of government. [That becomes] where all the action is. Except that's not where all the action is. The action is in West Oakland or East Oakland or Ashland or Cherryland [All parts of Alameda County, CA that are predominantly low-income and of color. West and East Oakland are urban, while Ashland and Cherryland are unincorporated]. That's where the action is or isn't. All of the money and time we waste supporting these big institutions that are all policy this and that, it's just a total waste.

**Navina**: I totally agree that the most important work is the work that's happening on the ground. And simultaneously, there are policies that are making it really hard for people to live in the ways that we want everyone to be able to live. I think the piece that's missing is exactly that—there are communities who are really hurt by the current system that are working to change it, and those stories need to be heard, because that's what will help move our culture and our movement forward.

**Hank**: [The movement is] emergent from healthy practices. It's not something you can design or construct. You can't program someone to be healthy. You can create healthy opportunities and experiences for them and they will become healthy on their own. Health emerges within a context of family, neighborhood and community that are working in a healthy way.

## Insights and Analysis

Hank and Navina offer a number of reasons that nutrition needs to be done differently if it is to matter to the youth and communities they work with. As organizers, they are certainly motivated in part by issues of health and health disparities, particularly when they are as glaring as the statistics Navina cited above. But in low-income communities of color, health and nutrition are inextricably linked to larger structural, cultural and economic inequalities. Navina and Hank offer two examples of what this more holistic approach can look like.

One area in which the disconnect between nutrition interventions and the lived experiences of low-income communities can be seen is language. Both Navina and Hank assert that the language in which nutritionists and public health officials describe their goals fails to resonate with the youth they work with. Sometimes this is clearly about cultural insensitivity and a lack of recognition of how racial difference can affect food choices. For Hank, this is exemplified by the nutritionist in Rochester who failed to understand why Mexican Americans continued to use lard. Navina similarly recalls her own preference for white rice over brown, claiming that the "cultural aspect of things is never part of the conversation." On an institutional level, this disregard for cultural preferences leads food and nutrition programs to prescribe items like enriched spaghetti, margarine and the iconic government cheese. Not only are the nutritional benefits of such items highly questionable, but more importantly, they represent the state's embrace of the industrial food system as the solution to health and nutrition problems.

The need for food and nutrition interventions to be "culturally relevant" has become almost ubiquitous, and even some government programs have begun to supplement products like those listed above with fresh fruit and vegetables. But Hank and Navina's conversation implies that, despite increasing nods to this concept, nutrition initiatives continue to regard food traditions, particularly those of communities of color, as targets for transformation. In contrast, Hank and Navina position a variety of indigenous diets as providing a multivocal alternative to the idea that everyone, regardless of their background, should desire the same kind of cuisine.

But relevance, for Navina and Hank, is about more than cultural background. As Hank said above, "We can't use those messages that are the official prescribed messages from the official prescribed institutions. It has to come from what happens down on the ground." While Hank and Navina recognize that the health consequences of contemporary diets are dire, particularly in low-income communities of color, they recognize the need for deep cultural work in order to make health and nutrition relevant to the lived experiences of those they work with.

To this end, they are careful not to put down the food choices of those they work with, even when they include sodas and McDonald's. Instead, they provide broader education about, and a chance to change, the circumstances under which those choices are made. And Navina emphasizes the importance of an ongoing and deep relationship between those who work to change nutrition patterns and practices (including herself) and those whose choices she remains committed to changing. Attempts to *do nutrition differently* might benefit from such a broad understanding of cultural relevance.

Hank and Navina's divergence from dominant approaches to health and nutrition is not only about language or cultural relevance. For these activists, health and nutrition is important, but not their ultimate goal. Instead, it is a way to make wider interventions in the structural economic disparities that burden the places where they work.

Given this very different approach, it is not surprising that Navina and Hank both criticize dominant approaches to health and nutrition as too individual, and aim to craft more collective responses. For Navina, this means collective organizing to influence policy, making structural changes in the way the industrial food system is organized. Her work with Live Real takes specific aim at the 2012 farm bill. Imagine how different the nutrition landscape would appear if the subsidies that make processing corn (and thus high fructose corn syrup) so cheap, were lessened or removed. What if organic produce was subsidized instead? This would mean a sweeping change in the kinds of foods that low-income people would have access to, giving them the opportunity to do nutrition differently without the need for targeted nutrition interventions. Rather than envision nutrition as a set of individual food choices, this approach locates nutrition in its social and even political context. Perhaps doing nutrition differently requires moving beyond targeted interventions and initiatives to participation in collective political work. Navina's comments about the need to reorganize the way that WIC money can be spent provides a more short-term example of how this approach could be engaged.

Hank, on the other hand, takes a different kind of collective approach through the creation of local economic initiatives. In the context of Dig Deep Farms, youth employees learn from and depend on one another. Not only do they gain skills around food cultivation, but they also learn public speaking and economic development. Hank suggests that the creation of economic stability in these communities, through well-paying jobs and local ownership, is the most promising way to address not only health but all kinds of disparities. This also fits with Navina's assertion that just importing a grocery store into a "food desert" neighborhood doesn't really affect nutritional habits if it isn't accompanied by economic development. After all, what good is proximity to a variety of food choices if you cannot afford them?

Hank's advice to those who seek to do nutrition differently might be to follow the example of the UC Davis researcher whose questions about food insecurity fell flat. Realizing that jobs and economic development were much more important to those she was studying, this researcher shifted her focus. This is not to say that a focus on health and nutrition need to go away entirely, but that it needs to be integrated with the immediate needs of those who are most harmed by the industrial food system. After all, it is clear from this conversation that issues of health and nutrition continue to motivate Navina and Hank's work. But integrating health with economic development can create a more holistic approach in which the food system is rightly seen as deeply connected to other systems, including those of racial and economic inequality and the uneven development that leaves residents of low-income communities of color without access to basic needs. This approach is not only about satisfying peoples' food and nutrition needs, but about making the structural changes necessary for people not to be in need to begin with.

Lastly, Hank and Navina do not target the eating habits of low-income people of color as something that needs to change, but instead work to inspire these people to become change-makers. Instead of shifting individual behavior, their kinds of approaches work to shift the food landscape, empowering those who are most harmed by the industrial food system to make the kinds of changes that are meaningful for them.

## References

Alkon, Alison Hope and Agyeman, Julian. 2011. *Cultivating Food Justice: Race, Class and Sustainability.* Cambridge, MA: MIT Press.

Alkon, Alison Hope and Christie Grace McCullen. 2010. Whiteness and Farmers Markets: Performances, Perpetuations … Contestations? *Antipode, a Radical Journal of Geography*, 43(4) 973-959.

Belasco, Warren J. 1993. *Appetite for Change: How the Counterculture Took on the Food Industry*. Ithaca, NY: Cornell University Press.

Guthman, Julie. 2008. If They Only Knew: Colorblindness and Universalism in Alternative Food Institutions. *The Professional Geographer*, 60(3): 387-397.

——. 2004. *Agrarian Dreams: The Paradox of Organic Farming in California.* Berkeley, CA: UC Press.

Harper, A. Breeze. 2011. Vegans of Color, Racialized Embodiment, and Problematics of the "Exotic." pp. 221-238 in Alkon, Alison Hope and Julian Agyeman (eds). *Cultivating Food Justice: Race, Class and Sustainability.* Cambridge, MA: MIT Press.

Just Food. "Food Justice." http://www.justfood.org/food-justice (accessed 7/7/10).

McClintock, Nathan. 2011. From Industrial Garden to Food Desert: Demarcated Devaluation in the Flatlands of Oakland, California. pp. 89-120 in Alkon, Alison Hope and Julian Agyeman (eds). *Cultivating Food Justice: Race, Class and Sustainability.* Cambridge, MA: MIT Press.

Minkoff-Zern, Laura-Anne, Nancy Pelluso, Jennifer Saurwhite and Christy Getz. 2011. Race and Regulation: Asian Immigrants in California Agriculture. pp. 65-86 in Alkon, Alison Hope and Julian Agyeman (eds). *Cultivating Food Justice: Race, Class and Sustainability*. Cambridge, MA: MIT Press.

Organic Trade Association. 2010. Industry Statistics and Projected Growth. http://www.ota.com/organic/mt/business.html.

United States Department of Agriculture. 2006. USDA Farmers Markets. http://www.ams.usda.gov/farmersmarkets (accessed 12/23/2007).

White, Monica Marie. 2010. Shouldering Responsibility for the Delivery of Human Rights: A Case Study of the D-Town Farmers of Detroit. *Race/ Ethnicity: Multidisciplinary Global Perspectives*, 3(2): 189-211.

# Thematic Tabs for Chapter 2

| Body |
| Discourse |
| Emotion |
| Race |
| Women |

Chapter 2

# Our Plates are Full:
# Black Women and the Weight of Being Strong

Tamara Beauboeuf-Lafontant

**Editors' Note:** This chapter engages the themes of *body, discourse, emotion, race,* and *women.* The chapter focuses on the notion of the 'strong' Black woman and the diverse ways in which Black women respond *with food* to this racialized feminine construct. In particular, the chapter portrays how overeating can become the outward expression of emotional states that have no direct mode of expression. Herein, the chapter offers a model for analyzing how the food-body relationships of ethnically and racially diverse women are shaped by particular cultural constructions of femininity, which hegemonic nutrition fails to engage.

## Introduction

Within the obesity epidemic afoot in the United States, Black women are a conundrum for health practitioners and researchers. Currently, about two-thirds of Black women age 18 and older are considered overweight, as compared to 60 percent of Latino populations, 47 percent of white women, and 25 percent of Asian/Pacific Islander women (Leigh and Huff 2006: 74). The pattern remains with regard to obesity: 35 percent of Black women, 26 percent of Latinos, 20 percent of white women, and 6.2 percent of Asian women meet the diagnostic criteria of having a Body Mass Index of 30 or above (Leigh and Huff 2006: 73). The tendency for Black women to weigh more than women from other ethnic and racial groups also carries across social class (Rand and Kuldau 1990, Williams 2002). Furthermore, such excess weight places Black women at elevated risk for chronic and debilitating conditions such as hypertension, heart disease, and diabetes, conditions which leave them with the shortest life expectancies among women and more years in ailing health (Fitzgibbon et al. 2008, Hill 2009).

Despite being among the most overweight and obese in society, Black women tend not to exhibit the discontent, diet-altering practices, and increases in weight-reducing physical activity observed among women from other racial and ethnic groups (Kwan 2010, Lovejoy 2001, Schuler et al. 2008). As researchers and the lay public puzzle over Black women's excess weight, they tend to place much attention on culturally inflected nutritional practices and choices (see, for example, Gans et al. 2009, Fitzgibbon et al. 2008). Scrutinized and generally found problematic has been the size of portions, choice of high-fat, high-sugar "soul" foods, and insufficient physical activity to offset such intake.

Although concerned with the health implications of overweight, former *Essence* magazine editor-in-chief Diane Weathers frames the question in novel and I believe instructive terms. Rather than focus narrowly on Black women's eating choices, Weathers asks, "What is it about being Black and female in America in 2003 that is causing an increasing number of us to carry more weight than we can handle?" (Weathers 2003a: 28). Her query recognizes the conceptual need for understanding racialized gender – how every expression of femininity is embedded in ideas about a woman's "race," and how this combination creates important variations in racially diverse women's experiences of their femininity. Ultimately the query demands that we take seriously the interconnections between racialized gender and food and nutrition practice in our attempt to expand upon and 'do nutrition differently.'

## Embodying Constructed Womanhoods

Much of feminine gender is a construction that serves the interests of women's subordination to masculinity as well as their adherence to racial and class hierarchies. Feminist work over the last 40 years has revealed how femininity is a performance that draws liberally and even exhaustively on women's physical and mental resources in order to render them intelligible as 'not men' in their social contexts. As a visual and visceral form, racialized and classed femininities are "an artifice, an achievement ... 'styles of the flesh,'" a set of manipulations rendered 'natural' through extensive self-surveillance and the alteration of one's appearance and demeanor (Bartky 1990: 65, see also, Gimlin 2002, Ussher 2004).

Paralleling such management of the physical body are the psychological accommodations of femininity. In everyday interactions, adolescent girls and women are encouraged to "leave yourself" and normalize self-silencing in order to effect socially acceptable forms of femininity (Anderson-Fye 2003). Depression, eating problems, and abuse in relationships have all been tied to the self-neglect and repression of authentic emotions outside of the "tyranny of nice and kind" (Brown and Gilligan 1992) that defines many femininities. While the toll of such feminine goodness is in the main of feminist inquiry, too little attention has been paid to how ethnically and racially diverse women respond in body and mind to cultural constructions of their goodness. Thus, we are limited in our understanding of how the "costly performance" (Beauboeuf-Lafontant 2009) of gender is experienced and borne on the bodies of all women. As I suggest in this chapter, the stress and emotional eating that Black women readily acknowledge in these pages are fundamentally tied to the life conditions placed upon them to be 'strong' Black women, the backbones of their communities and the ones left standing when all and everyone else has fallen.

## Being Strong: A Weighty Expectation

Among African Americans, strength defines adult feminine goodness. To be a strong Black woman is to demonstrate self-reliance, resilience, unselfishness, and unwavering race loyalty. For the ways it appears to have helped them manage material deprivation and political subordination from slavery to the present, strength is also a quality Black women proudly embrace as an unassailable virtue that reflects cultural authenticity and distinctiveness (Beauboeuf-Lafontant 2009).

Despite these positive associations, scholarship over the last 30 years has raised serious concerns about strength in regard to Black women's physical and mental health (Beauboeuf-Lafontant 2003, 2005, 2007, 2008, 2009, Boyd 1998, Edge and Rogers 2005, Gillespie [1978]1984, Hill 2009, Jones and Shorter-Gooden 2003, King and Ferguson 2006, Morgan 1999, Scott 1991, Wallace [1978]1990, Woods-Giscombe 2010). A growing Black feminist literature – which has been autobiographical, theoretical, clinical, and more recently empirical – regards strength as a set of exacting "accountability pressures" (Yancey Martin 2003: 358) that Black women encounter at home, in heterosexual romantic partnerships, at work, and in their community institutions. Growing up, girls are groomed into becoming strong Black women. As such, they learn "you're everything to everybody … and you don't think about yourself" (Carlisle Duncan and Robinson 2004: 91). In a study of Black women with post-partum depression, many interviewees identified with the concept of strength, which they described as the imperative to "keep on going," "get over it," "pick yourself up,"… "snap out of it," "just go on," "do what you are supposed to do," and "handle your problems" (Clark Amankwaa 2003: 310). As a lived experience, being strong entails exhibiting a stance of "pleasing the masses" (Black and Peacock 2011), engaging in "self-sacrificial" (King and Ferguson 2006) acts of caretaking, and supporting such other-directedness through ongoing self-neglect (Warren-Findlow 2006). Summarizes clinical social worker Terrie Williams (2008: 31), "We have embraced very destructive beliefs about our ability to 'handle it all,' our power to overcome in the face of trauma, our ability to put ourselves aside as we tend to the needs of our employers, partners, children, family – everyone but ourselves! So many of the Black women I meet live in terror of letting anyone down, but could care less about the number of promises they break to themselves every day."

Further ensconcing strength as the default definition of goodness is its comparative and deeply evaluative subtext. Strong Black women are measured against cultural and family icons, ranging from Sojourner Truth and Rosa Parks to female kin revered for their life wisdom and overcoming of adversity (McGee 2005: 10). Because enslavement is the reference point for adversity, contemporary Black women are rarely viewed as having legitimate grievances. As a result, writer Meri Nana-Ama Danquah (1998: 21) recalls that kin and friends were resistant to acknowledging her experiences of clinical depression. In their view, as an educated Black woman, she had little "to be depressed about." After all, "If our people could make it through slavery, we can make it through anything."

To achieve the stature of a strong Black woman, then, much self-neglect and emotional repression is required, and this has implications for Black women's health and eating. Poet-activist June Jordan ([1983]2000: xxix) presciently observed that the expectation that strong Black women will always make "a way outa no way" is not only "too much to ask/Too much of a task for any one woman," but that to accomplish this, a woman has to take "flesh outa flesh." Critical work in health reveals that Black women suffer in ways that their exceptionality denies.

### A Good Woman's Appetite: More Than Nutrition

Along with other racial groups, contradictions and tensions in feminine role shape Black women's relationship to food. The title of Becky Thompson's (1994) seminal study of eating problems among ethnically and racially diverse women – *A Hunger So Wide and So Deep* – immediately signals the important and overlooked dimension of appetite with regard to women's eating decisions. It recognizes that food is a literal and more importantly a symbolic part of all women's lives that allows them to manage experiences of trauma, powerlessness, and ambivalence with regard to reigning constructions of their goodness. Whether anorexia, bulimia, or compulsive overeating, eating problems operate on the level of literal nutrition as well as of metaphorical appetite. Each is a "transference process," through which women use their bodies to manage injustice in their lives (Thompson 1996).

For many 'strong' Black women, overeating is the outward expression of emotional states that have no direct mode of expression. This "emotional eating" (Thomas et al. 2008) is very much a quick fix (Robinson and Ward 1991), or a survival strategy that allows the user to temporarily evade pain, prolong pleasure, and engage in short-term self-care (Scott 1991, 10). It is a respite to which strong Black women gravitate, particularly because, as health activist Beverly Smith observes, "the emotional well-being of Black women [both within their communities and in the larger society] is pretty much ignored. The condition of our psyches, the inner lives, thoughts and feelings of Black women are not paid much attention to" (Lewis 1994: 178). Under such circumstances, food easily becomes a tool, a culturally available resource, for coping with the lack of public recognition that Black women – whom Zora Neale Hurston (1937: 14) famously characterized as 'the mule uh de world' – are fully human beings and entitled to lives beyond extensive service to others. Viewed as invulnerable to harm, Black women are actively impeded from recognizing and acting on hurts, disappointments, and fears more easily associated with other race-gender groups. As a result, their eating and weight gain become meaning-laden attempts to acknowledge and speak the realities that appeals to their strength systematically deny.

The painful irony of weight gain is that it, too, is easily subsumed by the strength discourse as a sign of a woman who can "handle the rough times better" (Allan, Mayo, and Michel 1993: 329). Furthermore, cultural norms that value meals as an expression of caring and affiliation, the aesthetic preference for thicker rather than thin figures, a view of dieting and weight-controlling exercise as suspiciously white endeavors (Carlisle Duncan and Robinson 2004, Parker et al. 1995), and beauty ideals more concerned with skin color and hair texture than weight (Hesse-Biber et al. 2004, Leeds 1994, Okazawa-Rey, Robinson and Ward 1987) collude to make overeating a coping strategy that can evade critique. Whereas clothing and hair are considered important body projects reflecting good grooming, self-pride, and individuality, weight is often framed as a relatively stable and a tacitly distinctive feature of Black women. Thus, the size and shape of a Black woman's body are often interpreted – within Black communities and in the larger society – as symbolic and immutable markers of both her degree of authenticity and strength. And because eating is not automatically problematized as it is in white communities, it is a safe strategy for registering discontent. It enables a woman to experience a break from her other-directedness, while not placing her under the direct scrutiny of those who hold her in high esteem for appearing strong. However, utilizing this culturally approved outlet means that "Instead of crying or dealing with our anger, depression, and pain," Black women can easily "suppres[s these emotions] with food" (Powers 1989: 134, 136, see also, Weathers 2003b: 190). As I suggest in the following sections, Black women's proscribed angers lead to being fed up and eating in anger, while their vulnerabilities drive self-protective attempts to eat themselves numb.

## I'm Fed Up

The exceptionality strength attributes to Black women paves the way for their exploitation – their profound service rendered to men, children, extended family, community institutions, and workplaces. Because a good Black woman is expected to "suppres[s] our pain to reduce someone else's" (Welsh 1979: 39), for many Black women, the emotional prohibitions of strength inform their eating.

Within a focus group, college students Kira[1] and Macy disclose how their overeating is associated with experiencing anger they can no longer deny.

> Kira: I feel like food is the easiest thing to get to, you know. . . . Let's say I'm having trouble, problems with people in general. Like, I go to work, I go to school every day, and people are always interrogating me or whatever. And it's like, "Well, right now, food will solve the problem. It'll satisfy me." You know, it's the easiest thing to get to.

> Macy: Shuts you up [said almost inaudibly].

---

1   All names are pseudonyms. Methodological information about interview data can be found in Beauboeuf (2009).

TB: What did you say?

Macy: Shuts you up for a second.

TB: Well, if we didn't, if you weren't "shut up for a second," what might happen?

Macy: You would explode. You would just start telling people off, and wouldn't care if you hurt anybody's feelings… I always keep stuff *in*, and I'll let it stick to me. And if you hold stuff in, it's going to eat away at you. Just like food, [it] will eat away at you, if you hold it in.

When Macy uses food to "shu[t] you up," she does so to keep from "exploding," that is, revealing those emotions she experiences but which a good Black woman cannot show. More critical than the fact of her eating is the motivated use of her body to absorb rather than express emotions and critiques of the social relations she finds unjust. The attendant weight gain she and other women experience can be understood, both as a kind of release and as a problematic reabsorption of feelings within the economy of their bodies. Venting through eating is inadequate, not simply because a woman's anger is not voiced directly to others, but also because overweight on Black women is often viewed as a physical marker of their strength or their ability to endure (Baturka, Hornsby, and Schorling 2000, Hebl and Heatherton 1997, Townsend Gilkes 2001).

Such overeating reflects the fact that strong Black women have few socially acceptable outlets for voicing a range of human emotions. Explains Michelle, a 21-year-old college student and teacher's aide, a fundamental tension exists: As a strong Black woman, one must "not be seen coping, basically," and yet, as a human, "you *do*, we *have* to cope, we *have* to cope with whatever we do." In a similar vein, Angie, a 24-year-old college student, challenges a common logic that focuses on excess weight as the source of Black women's ill health. Instead, she identifies the strains in women's lives and their lack of coping skills as the often hidden but very critical causes of both the overeating and the physical ailments that ensue. She argues, "It's not because we overweight that we got heart problems, and heart failures, and high blood pressure. It's 'cause we *stressed* out! And we don't know how to *vent* it the right way. Or, if we vent it, we vent it with a Häagen-Dazs ice cream or some other ice cream." However, although eating is socially safe, it is ultimately a flawed outlet for growing wells of frustration.

Deep in our interview, Michelle recounts a binge episode occasioned by the dynamics of her racialized workplace.

Let's say that you, you're eating food. And let's say that you're in the lounge or something, and you're eating amongst white people. You're *ticked off* at what just happened, you know. That the fact of what [a supervisor] just did to me, you know, just didn't allow me to go. So, okay, I went to [a donut shop], and instead of getting two donuts, I got *six* donuts. And I'm eating them one after *the other*.

And I have donut holes, and then my *lunch*, and then *ice cream*. So, white people will look at you, or any other type of race who *don't* understand what you're going through, will look at you and go, 'What the hell's her problem? Why are you eating?' One might ask, 'Michelle, that's not really good for you.' But they don't ask, 'Michelle, what's wrong?' Because it's *obvious* that something has to be wrong with you. But the way you cope, it's not obvious to you that the way you cope is different from the way I cope.

Michelle's plate is literally full. Her choice of foods, amounts consumed, the relationship of intake to expenditure are troubling, particularly as she has been recently diagnosed as clinically obese by her doctor and is in need of losing 100 pounds. She is at an unhealthy weight early in her adult life, and while she does not have the attendant health problems, if maintained, her weight and inactivity will set her on a culturally familiar and some might say epidemic path of ill-health. However, focusing solely on these "bad" foods – Michelle's nutrition – can easily overlook the fact that her eating is a coping tool for anger she cannot openly express at work. Although she is cognizant of the fact that others – particularly racial outsiders – will acknowledge and even comment on the poor nutritional value of her meal, she asserts something more basic is amiss: that she was mistreated in her place of work, and exploited as a person of lower social standing and limited power.

As with many other employed Black women, Michelle is "mammified" or pressed to assume a status-reassuring deference to whites, particularly in workplaces. Mammification (Omolade 1994) is a deeply ingrained manifestation of white privilege in interracial, specifically black/white interactions. Very intentionally, mammification invokes the long history of racialized and gendered comfort imagined in the person of a large, African-featured Black woman (McElya 2007). Explains cultural critic bell hooks (1991: 154):

> [R]acist and sexist assumptions that Black women are somehow 'innately' more capable of caring for others continues to permeate cultural thinking about Black female roles. As a consequence, Black women in all walks of life, from corporate professionals and university professors to service workers, complain that colleagues, co-workers, supervisors, etc. ask them to assume multi-purpose caretaker roles, be their guidance counselors, nannies, therapists, priests; i.e., to be that all-nurturing "breast" – to be the mammy.

Mammified Black women may receive verbal praise, but rarely does this recognition materialize as tangible rewards of power, choice, or respect in the workplace. The work extorted from Black women read as workplace mammies is very significant, leading to a "role strain by enforcing the belief that Black women happily seek multiple roles rather than assuming them out of necessity, that they effortlessly meet their many obligations, and that they have no desire to delegate responsibilities to others" (West 2008: 290). Michelle explains that having to work amidst these assumptions that her job is to "tak[e] care of [the teacher's] problems before I'm taking care of myself" is a dehumanizing experience: "It's like if somebody is trying to put you in a box and put a lid over it, and only give you enough air to breathe, you know ... But not let people see who's in the box ...". Michelle is eating the anger she feels incapable of expressing to a white supervisor, since she is expected to be strong – that is, devoted to her co-workers and devoid of any discontent with workplace decisions and dynamics that are harmful to her.

A 47-year-old factory worker, Marva similarly questions whose interests are served by her strength. She insists that workplace attributions of her strength are in effect a "backhanded compliment" that initially "tickles your ears" until she realized that her co-workers were using this praise to "dump, and dump, and dump" work on her that white women employees were seen as being inherently "unable" to handle. Like Michelle, she sees her options as limited. "If I whined and threw a big fit, I probably wouldn't have a *job*," an effective recourse open to her white co-workers. Both Michelle and Marva know that complaining among Black women in majority-white workplaces – where they are expected to be mammy, an eternally caring and providing figure with no needs of her own – runs the risk of being dismissed as insubordination, as the manifestation of an inappropriately "angry Black woman" (Harvey Wingfield 2007).

**Eating Myself Numb**

For Black women laboring under the burden of keeping up the appearance of their unflappability while immersed in social conditions that assault their minds, bodies, and spirits, binge eating can bring them much-needed respite and protection. As Sharlene Hesse-Biber (1997: 111) finds, "The intake of large quantities of food in a short time period can serve to numb, soothe, and literally 'shield' (with fat) some women from physical and emotional trauma."

The outlet of eating is particularly necessary for women who are torn between their human need to cope and the mandate of strength, which insists that Black women are above any assistance or concern and can manage all. Bingeing can tamp down the harm, disappointment, or outrage that Black women are not supposed to experience. As expressed by a college graduate reflecting on her compulsive eating (Johnson 2005: 197):

> I was completely unprepared when the job market didn't stand up and take notice of my studies abroad and expensive diploma. Soon I was relying on my mother, stepfather, and boyfriend for financial support. I couldn't even pay my rent. What I *could* do was eat myself numb. I consumed unspeakable amounts of whatever I could buy in the vicinity of my downtown Brooklyn studio apartment. French fries. Pizza. Macaroni and cheese. Fried whiting sandwiches shellacked with hot and tartar sauce. Food became a drug I'd use when I was feeling happy,

sad, or somewhere in between. It was a reward. It was a sedative. Food was my companion ... "[E]at[ing] myself numb" is a purposeful tool; massive food intake serves as "a reward ... a sedative ... my companion," a way of meeting needs for recognition, support, and protection (Johnson 2005: 197).

As persons who cannot suffer, strong Black women are denied a vocabulary for examining their overwhelming obligations to others, their limited resources for meeting those demands, and their needs for and support. As an activity, eating enables Black women to register and attend to some of their needs without disrupting the fiction of their strength. Jennifer, a 29-year-old bank employee and divorced mother, views the eating of the strong and large women in her family as allowing them what strength does not – time away from "the struggles of life. You know, the one time in the world they have to rest is, 'Hey, eat.' One thing they know they can do well is eat and that's their quiet time. You eat and you relax, you know." Amid these relational and economic conditions, "food [easily becomes] a vehicle that is used to comfort us when we may not have much else" (Walcott-McQuigg et al. 1995: 512).

Recent and growing research additionally suggests that overweight can offer protection and an "anonymity" (Mitchell and Mitchell 2004: 23) particularly important to victims of abuse. Fat desexualizes, a point not lost on survivors of sexual abuse. Black women experience higher levels and often more extreme forms of intraracial violence across their lifespan than women from other racial/ethnic groups (Nicolaidis et al. 2010). Furthermore, codes of racial and cultural loyalty are typically predicated on a gender silence that compels 'good' Black women to minimize and hide the often severe abuse they experience in their intimate relationships with men (Tarrezz Nash 2005, see also, Bell 2004, Weathers 2003c: 161). As a result, Black women victims often feel compelled to "display inner strength and minimize the impact" of their violations, leaving the erroneous "impression that Black women are relatively unscathed by their sexual trauma" (West 2006: 6). Under such conditions of violence and silence, women may have much motivation to "downplay their femininity" (experienced as vulnerability) through weight gain (McGee 2005: 62, see also, Thompson 1994).

Binge eating disorder is more common among the obese, particularly if there is a history of childhood sexual abuse (Stevelos and White nd), and in a recent community survey, clinical levels of recurrent binge eating were found to be more common among Black women than white women (Striegel-Moore et al. 2000). As binge-eating disorder gains traction as a diagnostic category and eating problem, the protective function of overweight becomes apparent. Yasmin, a 32-year-old educator, observes from her own experiences of sexual abuse that fat can effectively "lock" the body in an attempt to protect it from further heterosexual attention and violation.

I will not go into detail, but I am an abuse survivor. I think a lot of women, who were taught that you should not have sex before marriage, are *hiding* the sexy.

> And I think there's a whole culture, and a whole way of *covering up with fat,*
> things that you are not supposed to be using … Sexual abuse is a big problem
> in the Black community. If *you* had somebody hurt your physical body, the one
> thing you're going to do, or you're *not* going to do, is sort of let it all out … And
> I think there are a lot of obese women who are basically hiding their bodies.

Yasmin recalls periods during her adolescence of engaging in "disordered,
binge eating. Just out of stress, you know, where I took a 15 minute period" to
ingest food. Her weight gain was an intentional way of "lock[ing]" her body up so
that neither she nor anyone could use and expose it to further abuse. Remarking
on the existence of violence in the lives of women she knows, Tamika, a 40-year-
old public health officer, also understands that "weight is a cover for some deeper
issues" which a woman either "hasn't dealt with" or that others are not willing to
acknowledge:

> You know, it's easier to talk about weight than to talk about, you know, 'I'm in
> an abusive relationship.' Because people can accept and be comfortable talking
> to you about weight, but when you talk about, you know, 'My husband's beating
> me everyday,' people are going to *push away* from you.

Eating can function as a protective measure for many Black women. It is
a backstage behavior that allows them to keep up their "game face" and avoid
what a blogger reveals is "simply too shameful to say," that is, "'Help me. I'm
drowning'" (Black and Peacock 2011, 147). Because overeating and attendant
weight gain can provide emotional and physical "layers of protection" to hide
the self, in "shedding layers" explains Yasmin one "sheds protection." Thus, to
lose weight would be to give up defenses that have served such women well in
unpredictable and hurtful circumstances.

## In Failing Health: Weighed Down by Strength

Even when faced with dire health conditions, strong Black women struggle with
a greater lifestyle change – that is, "being 'selfish' about [their] health" (Weaver,
Gaines, and Ebron 2000: 134). Concern about weight seems a "luxury" and an
indulgence they cannot afford, especially when there are so many other pressing
needs to minister to others, and when their goodness and authenticity are riding on
their selflessness and other-directedness.

Speaking about a woman she knows, whose overweight now has impending
fatal consequences, 33-year-old Tasha invokes the phrase "let[ting] herself go."
Typically a condemnation of women who fail to take adequate steps to guard against
stigmatized weight gain, the phrase seems to have a more profound significance in
this example. As Tasha describes, this woman is "unhealthy … sick" and has been
told by her doctor "that if she doesn't lose 50 pounds, she is going to die." The

fact that she has not made dietary and lifestyle changes strikes Tasha as a form of suicide: "Why do you have to kill yourself, when there's plenty [of] other outside things that'll do it, you know?" Although critical of this woman's current nutrition, Tasha speaks of the admiration she has for her strength.

> She raised her two girls and her son by herself ... Actually, she's a woman who encourages me. So, you know, I mean, she knows about everything that happened in the family 'cause she's somebody who I could go and talk to, and she encourages me that, you know, 'Don't let that discourage you, you know. Keep on going and everything like that.' But I wouldn't, if I *disqualified* her as a [strong] woman, it would be because of what she is letting *herself* go through. Not because of what she does for her family. Because she's still there for them and all of that, but, I mean, I would want her to do better by *herself.*

The weight that this woman has gained in recent years demonstrates to Tasha that something is amiss in her life. "Do[ing] better by *herself*" would entail following the advice of her doctor and making her health a priority.

Arguably what Tasha and many observers of Black women do not recognize is that compliance with the doctor's orders is a challenge for any strong Black woman. What such a woman knows too well is how to put more stock in the image of her strength than in her own emotional reality. Writes journalist Jill Nelson (1995: 68), "So much of our conditioning has been to be the strong figure in the family – the backbone, the one who can take more weight than anyone – that it is hard to know when we're overloaded." Similarly writer Rosemary Bray (1992: 54) speaks of her obesity as the embodiment of the "emotional weight" she carries specifically as a Black woman:

> We are forever working, loving, volunteering, scolding, nurturing and organizing – but nearly always for others ... Black women have assumed so much responsibility in this culture I often wonder how we can still stand up. Who and what supports us? In truth, it is most likely food that sustains us ...

Bray correctly concludes that her hunger is a social problem and that like other strong Black women, she is "immensely hungry for much more than food ... for the things all of us are really hungry for: hungry to be truly seen and known, hungry to be accepted the way I am. There may be no more difficult desire for an African-American woman to fulfill" (Bray 1992: 90).

Ironically, Black women's compensatory eating and attendant weight gain feed into the construction of them as operating with distinctive emotional and physical capacities. Such exceptionality attributed to Black women may help explain sociologist Cheryl Townsend-Gilkes' (2001: 183) observation that while "the most respected physical image of black women, within and outside of the community, is that of the large woman ... some of the most powerless women in the community struggle with overweight and its unhealthy consequences." Thus, with regard to their physical health and nutrition, the concern we develop from a sensitivity to the discourse of strength is that the weight-related diseases that plague the Black female community may be embodied manifestations of the contradictory distinction of being strong and "last on every shopping list in town," including one's own (Wallace 1990: 227). Women who rely on overeating to manage disparities between what they are able to do and what strength dictates they should manage are also bound by the belief that anything other than persuasive shows of emotional restraint, race loyalty, self-reliance, and invulnerability "becomes a flaw in who we are," rather than a reflection of problematic social conditions in their lives. In this way, overeating is a costly form of self-silencing that allows Black women to stave off the direct expression of emotions that would call much of their social worlds into question.

From years of "picking up" strength and later submerging her discontent through compulsive overeating, Traci now recognizes that although "[strength's] the way you're being taught," health and wellness remain elusive until "you … grow out of it. Or eventually, as a grown woman, find something that works for you." Traci's loss of 60 pounds and avoidance of Type-2 diabetes required that she see and squarely question the depleting service of strength:

> Years ago, I want to make sure *everybody* was happy. There's no way, *in the world*, you can ever make sure everybody is happy. And you try to play that superwoman role, where you *try* your best … And that's just not how life goes [chuckle]. I mean, sincerely, it's *not* how life goes, and I found that out.

She credits a daily walking routine not simply with her weight loss but with liberating a subjectivity beyond strength's self-silencing prescriptions. Particularly striking is how she describes "being free, out in the air, the wind blowing," the time alone allowing for a direct expression of her needs:

> And you *think*, and when you walk and think, then you try to figure out, 'Well, what's best for me right now? What type of woman do I want to be *right now?*' And then you start putting things into a perspective, because then I *divorced* … Yes, I started walking, and I said, 'This isn't. I'm *not* happy. And if I'm not happy, I don't have to portray this role anymore that for him I'm going to be happy.'… You have to make a decision to go out and help yourself, and then you're no longer thinking for just your daughter. You're thinking for *you*, as an African American woman. Because once you're dead and gone, nobody's going to know *what* happened, why you kept it inside.

Through walking, Traci was able to focus on her subjective appraisal of life, and not simply on her dependents, their needs, and their expectations of her. Being able to step behind these demands, she uncovered and regained a "voice" and an "identity" which she then used to question and distance herself from the role of strength: "I had time to think, and then, you know, deal with who I am. Who, basically, I am ... I took control over my life, then ... I started thinking, and I started becoming *bold* [chuckle], I started making decisions, and I'm talking *later* ... In my late 30s." Walking was an outlet that amplified rather than muted Traci's voice as a person much more than just strong.

## Moving Beyond Strength: A Weigh Out

Although expressed as a tribute and a virtue, the lived experiences of Black women reveal that strength consigns them to much silence, stoicism, and selflessness – a combination that leaves them with deep needs for the respite and protection which some meet through excessive eating. Overeating among strong Black women is fundamentally tied to the tension between being human but having to present oneself as strong.

Strong Black women's tendency to mask their emotions, frustrations, angers, and fears contributes to the weight they may carry through overeating, lack of regular exercise, or a general sense that focusing on their own needs is trivial or selfish. Nutrition discussions that approach Black women's overweight as fundamentally about portion size and the healthfulness of food choices are at best a superficial reflection of the appetites Black women have and that are routinely denied in their everyday interactions at home, in their communities, and in the public sphere. As the accounts in this chapter reveal, Black women hunger for support, assistance, and loads not so burdensome. The foregoing analysis suggests effective weight management for Black women will remain elusive as long as they are hailed as the mules of the world and compelled in the name of this strength to deny their humanity. Such women are "overworked, undervalued, and under pressure" (Williams 2008: 31), and resolve such role strain by further withholding their own needs or meeting them obliquely through eating practices.

Nutrition campaigns need to examine the weighty plates Black women carry as they are unfairly expected to be beyond harm, concern, and need. Their eating reflects material want and cultural food traditions, but also – and this is most often overlooked – everyday, ongoing interactions that pressure them to comport themselves as more capable and less vulnerable than their actual human resources and abilities allow.

## References

Allan, Janet, Kelly Mayo, and Yvonne Michel. 1993. Body size values of white and Black women. *Research in Nursing and Health*, 16(5), 323-333.

Anderson-Fye, Elieen P. 2003. Never leave yourself: Ethnopsychology as mediator of psychological globalization among Belizean schoolgirls. *Ethos*, 31(1), 59-94.

Bartky, Sandra Lee. 1990. *Femininity and Domination: Studies in the Phenomenology of Oppression*. New York: Routledge.

Baturka, Natalie, Paige P. Hornsby, and John B. Schorling. 2000. Clinical implications of body image among rural African-American women. *Journal of General Internal Medicine*, 15(4), 235-241.

Beauboeuf-Lafontant, Tamara. 2003. Strong and large Black women? Exploring relationships between deviant womanhood and weight. *Gender & Society*, 17(1), 111–121.

Beauboeuf-Lafontant, Tamara. 2005. Keeping up appearances, getting fed up: The embodiment of strength among African American women. *Meridians*, 5(2), 104-123.

Beauboeuf-Lafontant, Tamara. 2007. 'You have to show strength': An exploration of gender, race, and depression. *Gender & Society*, 21(1), 28–51.

Beauboeuf-Lafontant, Tamara. 2008. Listening past the lies that make us sick: A voice- centered analysis of strength and depression in Black women. *Qualitative Sociology*, 31(4), 391-406.

Beauboeuf-Lafontant, Tamara. 2009. *Behind the Mask of the Strong Black Woman: Voice and the Embodiment of a Costly Performance*. Philadelphia: Temple University Press.

Bell, Ella Louise. 2004. Myths, stereotypes, and realities of Black women. *The Journal of Applied Behavioral Science*, 40(2), 146-159.

Black, Angela Rose and Nadine Peacock. 2011. Pleasing the masses: Messages for daily life management in African American women's popular media sources. *American Journal of Public Health*, 101(1), 144-150.

Boyd, Julia A. 1998. *Can I Get a Witness?: For Sisters, When the Blues is More Than a Song*. New York: Dutton.

Bray, Rosemary. 1992. Heavy burden. *Essence*, 22(9), 52-54, 90-91.

Brown, Lyn Mikel and Carol Gilligan. 1992. *Meeting at the crossroads: Women's psychology and girls' development*. New York: Ballantine.

Carlisle Duncan, Margaret and T. Tavita Robinson. 2004. Obesity and body ideals in the media: Health and fitness practices of young African-American women, *Quest*, 56(1), 77-104.

Clark Amankwaa, Linda. 2003. Postpartum depression among African-American women. *Issues in Mental Health Nursing*, 24(3), 297-316.

Danquah, Meri Nana-Ama. 1998. *Willow Weep for Me: An African American Woman's Journey Through Depression*. New York: One World.

Edge, Dawn and Anne Rogers. 2005. Dealing with it: Black Caribbean women's response to adversity and psychological distress associated with pregnancy, childbirth, and early motherhood. *Social Science and Medicine*, 61(1), 15-25.

Fitzgibbon, Marian, Melinda R. Stolley, Linda Schiffer, Lisa K Sharp, Vicky Singh, and Alan Dyer. 2008. Obesity reduction Black intervention trail (ORBIT): Design and baseline characteristics. *Journal of Women's Health*, 17(7), 1099-1110.

Gans, Kim, Patricia M. Risica, Usree Kirtania, Alishia Jennings, Leslie O. Strolla, Matilda Steiner-Asiedu, Norma Hardy, and Thomas M. Lasater. 2009. Dietary behaviors and portion sizes of Black women who enrolled in *SisterTalk* and variation by demographic characteristics. *Journal of Nutrition Education and Behavior*, 41(1), 32-40.

Gillespie, Marcia Ann. [1978]1984. The myth of the strong Black woman. In *Feminist frameworks: Alternative theoretical accounts of the relations between women and men*, edited by Alison M. Jaggar and Paula S. Rothenberg. New York: McGraw-Hill, 32-35.

Gimlin, Debra. 2002. *Body work: Beauty and self-image in American culture*. Berkeley: University of California.

Harvey Wingfield, Adia. 2007. The modern mammy and the angry Black man: African American professionals' experiences with gendered racism in the workplace. *Race, Gender & Class*, 14(1/2), 196-212.

Hebl, Michelle and Todd Heatherton. 1997. The stigma of obesity in women: The difference is black and white. *Personality and Social Psychology Bulletin*, 24(4), 417-426.

Hesse-Biber, Sharlene. 1997. *Am I Thin Enough Yet? The Cult of Thinness and the Commercialization of Identity*. New York: Oxford University.

Hesse-Biber, Sharlene, Stephanie A. Howling, Patricia Leavy, and Meg Lovejoy. 2004. Racial identity and the development of body image issues among African American adolescent girls. *The Qualitative Report*, 9(1), 49-79.

Hill, Shirley A. 2009. Cultural images and the health of African American women. *Gender & Society*, 23(6), 733-746.

hooks, bell. 1991. Black women intellectuals. In *Breaking Bread: Insurgent Black Intellectual Life*, bell hooks and Cornel West. Boston, MA: South End Press, 147-164.

Johnson, Elon D. 2005. Miss Piggy gets depressed, too. In *Naked: Black Women Bare All About Their Skin, Hair, Lips, and Other Parts*, edited by Ayana Byrd and Akiba Solomon. New York: Perigee, 195-201.

Jones, Charisse and Kumea Shorter-Gooden. 2003. *Shifting: The Double Lives of Black Women in America*. New York: Harper Collins.

Jordan, June. [1983]2000. Oughta be a woman. In *Home Girls: A Black Feminist Anthology*, edited by Barbara Smith. New Brunswick, NJ: Rutgers University, xxix.

King, Toni C. and S. Alease Ferguson. 2006. 'Carrying our burden in the heat of the day': Mid-life self-sacrifice within the family circle among Black professional women. *Women & Therapy*, 29(1/2), 107-132.

Kwan, Samantha. 2010. Navigating public spaces: Gender, race, and body privilege in everyday life. *Feminist Formations*, 22(2), 144-166.

Leeds, Maxine. 1994. Young African-American women and the language of beauty. In *Ideals of Feminine Beauty: Philosophical, Social, and Cultural Dimensions*, edited by Karen Gallagher. Westport, CT: Greenwood, 147-159.

Leigh, Wilhelmina A. and Daniele Huff. 2006. Women of Color Health Data Book, 3rd edition [Online: Office of Research on Women's Health, National Institutes of Health, Washington, DC]. Available at: http://orwh.od.nih.gov/pubs/WomenofColor2006.pdf [accessed 7 April 2008].

Lewis, Andrea. 1994. Looking at the total picture: A conversation with health activist Beverly Smith. In *The Black Women's Health Book: Speaking for Ourselves, New Expanded Edition*, edited by Evelyn C. White. Seattle, WA: Seal Press, 172-181.

Lovejoy, Meg. 2001. Disturbances in the social body: Differences in body image and eating problems among African American and white women. *Gender & Society*, 15(2), 239-251.

McElya, Micki. 2007. *Clinging to Mammy: The Faithful Slave in Twentieth-Century America*. Cambridge, MA: Harvard.

McGee, Robyn. 2005. *Hungry For More: A Keeping-It-Real Guide for Black Women on Weight and Body Image*. Emeryville, CA: Seal Press.

Mitchell, Stacy Ann and Teri D. Mitchell. 2004. *Livin' Large: African American Sisters Confront Obesity*. Roscoe, IL: Hilton Publishing.

Morgan, Joan. 1999. *When Chickenheads Come Home to Roost: My Life as a Hip-Hop Feminist*. New York: Simon & Schuster.

Neale Hurston, Zora. 1937. *Their Eyes Were Watching God*. Greenwich, CT: Fawcett.

Nelson, Jill. 1995. Beyond the myth of the strong Black woman. *Heart & Soul*, 8 (January), 68.

Nicolaidis, Christina, Vanessa Timmons, Mary Jo Thomas, A. Star Waters, Stephanie Wahab, Angie Mejia, and S. Renee Mitchell. 2010. 'You don't go tell white people nothing': African American women's perspectives on the influence of violence and race on depression and depression care. *American Journal of Public Health*, 100(8), 1470-1476.

Okazawa-Rey, Margo, Tracy Robinson, and Janie Victoria Ward. 1987. Black women and the politics of skin color and hair. *Women & Therapy*, 6(1/2), 89-102.

Omolade, Barbara. 1994. *The Rising Song of African American Women*. New York: Routledge.

Parker, Sheila, Mimi Nichter, Mark Nichter, Nancy Vuckovic, Colette Sims, and Cheryl Ritenbaugh. 1995. Body image and weight concerns among African American and white adolescent females: Differences that make a difference. *Human Organization*, 54(2), 103-114.

Powers, Retha. 1989. Fat is a black women's issue. *Essence*, 20(6), 75, 78, 134, 136.

Rand, Colleen S.W. and John M. Kuldau. 1990. The epidemiology of obesity and self-defined weight problem in the general population: Gender, race, age, and social class. *International Journal of Eating Disorders*, 9(3), 329-343.

Robinson, Tracy and Janie V. Ward. 1991. A belief in self far greater than anyone's disbelief: Cultivating resistance among African American female adolescents. In *Women, girls and psychotherapy: Reframing resistance*, edited by Carol Gilligan, Annie G. Rogers, and Deborah L. Tolman. Binghamton, NY: The Haworth Press, 87-104.

Schuler, Petra B., Debra Vinci, Robert M. Isosaari, Steven F. Philipp, John Todorovich, Jane LP Roy, and Retta R. Evans. 2008. Body-shape perceptions and body mass index of older African American and European American women. *Journal of Cross Cultural Gerontology*, 23(3), 255-264.

Scott, Kesho Yvonne. 1991. *The Habit of Surviving*. New York: Ballantine.

Stevelos, JoAnn and Candace White. ND. Sexual abuse and obesity: What's the link? Part 1. [Online: Obesity Action Coalition]. Available at: http://www. obesityaction.org/magazine/ywm21/Sexual%20Abuse%20and%20Obesity. pdf [accessed: 21 February 2011].

Striegel-Moore, Ruth, Denise Wilfley, Kathleen Pike, Faith-Anne Dohm, and Christopher Fairburn. 2000. Recurrent binge eating in Black American women. *Archives of Family Medicine*, 9(1), 83-87.

Tarrezz Nash, Shondrah. 2005. Through Black eyes: African American women's constructions of their experiences with intimate male partner violence. *Violence Against Women,* 11(11), 1420-1440.

Thomas, Andria M., Ginger Moseley, Rayvelle Stallings, Gloria Nichols-English, and Peggy J. Wagner. 2008. Perceptions of obesity: Black and white differences. *Journal of Cultural Diversity*, 15(4), 174-180.

Thompson, Becky. 1994. *A Hunger So Wide and So Deep: American Women Speak Out on Eating Problems*. Minneapolis: University of Minnesota.

Thompson, Becky. 1996. 'A way outa no way': Eating problems among African American, Latina, and white women. In *Race, Class, and Gender: Common Bonds, Different Voices*, edited by Esther Ngan-Ling Chow, Doris Wilkinson, and Maxine Baca Zinn. Thousand Oaks, CA: Sage, 52-69.

Townsend Gilkes, Cheryl. 2001. *'If it Wasn't for the Women ...': Black Women's Experience and Womanist Culture in Church and Community*. Maryknoll, NY: Orbis.

Ussher, Jane. 2004. Premenstrual syndrome and self-policing: Ruptures in self-silencing leading to increased self-surveillance and blaming of the body. *Social Theory and Health*, 2(3), 254-272.

Walcott-McQuigg, Jacqueline A, Judith Sullivan, Alice Dan, and Barbara Logan. 1995. Psychosocial factors influencing weight control behavior of African American women. *Western Journal of Nursing Research*, 17(5), 502-20.

Wallace, Michele. [1978]1990. *Black Macho and the Myth of the Superwoman*. New York: Verso.

Wallace, Michele. 1990. *Invisibility Blues: From Pop to Theory*. New York: Verso.

Warren-Findlow, Jan. 2006. Weathering: Stress and heart disease in African American women living in Chicago. *Qualitative Health Research*, 16(2), 221-237.

Weathers, Diane. 2003a. Straight talk: Our bodies, ourselves. *Essence*, 34(5), 28.

Weathers, Diane. 2003b. Our bodies, ourselves. *Essence*, 34(5), 188-194.

Weathers, Diane. 2003c. Black America's dirty little secret. *Essence*, 34(3), 161-162.

Weaver, Roniece, Fabiola Gaines, and Angela Ebron. 2000. *Slim Down Sister: The African-American Women's Guide to Healthy, Permanent Weight Loss*. New York: Plume.

Welsh, Kariamu. 1979. I'm not that strong. *Essence*, 9(10), 39.

West, Carolyn M. 2006. Sexual violence in the lives of African American women: Risk, response, and resilience. [Online: Applied Research Forum, National Online Resource Center on Violence Against Women]. Available at: *http://new.vawnet.org/Assoc_Files_VAWnet/AR_SVAAWomen.pdf* [accessed: 21 February 2011].

West, Carolyn M. 2008. Mammy, Jezebel, Sapphire, and their Homegirls: Developing an "oppositional gaze" toward the images of Black women. In *Lectures on the psychology of women, 4th edition*, edited by J. Chrisler, C. Golden, and P. Rozee, New York: McGraw-Hill, 286-299.

Williams, David R. 2002. Racial/ethnic variations in women's health: The social embeddedness of health. *American Journal of Public Health*, 92(4), 588-597.

Williams, Terrie M. 2008. *Black Pain: It Just Looks Like We're Not Hurting*. New York: Scribner.

Woods-Giscombe, Cheryl L. 2010. Superwoman schema: African American women's views on stress, strength, and health. *Qualitative Health Research*, 20(5), 668-683.

Yancey Martin, Patricia. 2003. 'Said and done' versus 'saying and doing': Gendering practices, practicing gender at work. *Gender & Society*, 17(3), 342-366.

# Thematic Tabs for Chapter 3

| |
|---|
| *Access* |

| |
|---|
| *Colonial* |

| |
|---|
| *Discourse* |

| |
|---|
| *Nature* |

| |
|---|
| *Structure* |

| |
|---|
| *Women* |

Chapter 3

# Other Women's Gardens: Radical Homemaking and Public Performance of the Politics of Feeding

Kirsten Valentine Cadieux

**Editors' Note:** This chapter engages with the thematic tabs of access, colonial, discourse, nature, structure and women. The chapter's contributions to doing nutrition differently come through Cadieux's critical attentiveness to gardening stories – exploring how radical gardening stories may be useful for addressing food, feeding, and nutrition in a number of different ways. Furthermore, Cadieux shows how the performance of gardening and other food work provides opportunities to try out alternative navigations of the social tensions (and contentious politics) of household- and community-scale feeding practices. Cadieux emphasizes reflexivity concerning these tensions and politics and thus reinforces the call for reflexive engagement with numerous conflicts that have been highlighted in the critical literature on women's food and nutrition practices.

This chapter explores the current popularity of food gardening in terms of its implications for studying food, feeding, and nutrition. Food gardening holds a special place in many people's imagination of food security and good nutrition. Gardens are an easy-to-imagine source of fresh produce, which itself has become an icon of healthy eating. Access to space to garden is often taken-for-granted as an obvious way to improve people's diets.[1] A considerable history of environmental policy work to ensure access to gardens (Bassett 1981) has made garden land-use arrangements so easy for people to imagine, in fact, that gardens are often depicted as a form of self-sufficiency that is available for *anyone willing to put up effort*. Food gardens are seen to enable a kind of self-sufficiency that positions food gardeners outside the constraints of market conditions and power relations

---

1  This is especially the case from Anglo-American perspectives, where land is considered relatively abundant, and food gardens are seen as a highly legitimate land use. Many people imagine gardening as a default use of suburban and rural yards and gardening is also becoming increasingly visible in the form of urban allotments, community gardens, or rooftop or container gardening. This is significant in the context of current politics around urban gardening, in which food-related land uses are justified where market-based land use values would dictate against it.

that otherwise constrains food access (Astyk 2009, Hayes 2010, FAO 2000). At the same time, gardening appears to have been so successfully associated with "lifestyle"—e.g., in newspapers and magazines along with other uses for disposable income like fashion, travel, and gourmet cooking—that the *labor* of food gardening tends to be forgotten, downplaying the practical work involved in gardening in favor of highlighting meaningful enjoyment. In this chapter, I explore the stories and performances of food gardening as they relate to the politics of food system reform (often referred to as "feeding work" or "good food work"). I focus on public gardening sites that accentuate the sociality of food work, and allow us to see how gardening practices (that are often considered to be privately conducted at the household scale) are related to more public political contexts. Public sites especially exhibit attempts to assert the political value of garden projects—and also concern about these projects' limitations: even if garden work contributes something to current efforts to improve food systems and nutrition, how significant is this contribution, especially in the face of overwhelming systemic food system problems? I read the stories of gardeners with whom I have worked in extended case studies in Canada, the U.S., and Aotearoa New Zealand, and also of gardener authors, in terms of their efforts to represent the accomplishments of their gardening work.

In addition to the more obvious tasks of food production, gardening can involve engaging the social relations by which food systems are organized. Promoters of "civic agriculture" (or of critical versions of neo-agrarianism) argue that this engagement with the social and economic contexts of the food system is one of the more significant and legitimate justifications for food localization efforts (Lyson 2004, Cadieux 2005). Building on a long history of politically-oriented gardening and writing about food and feeding work (from radical back-to-the-land gardeners like Helen and Scott Nearing to radical food reform writers like Sally Fallon), a new generation of "radical homemakers," as Hayes calls them in her 2010 book about "reclaiming domesticity from a consumer culture," builds on a narrative tradition of political storytelling. Radical homemaker authors (and many readers of books like Sharon Astyk's *Nation of Farmers* or Barbara Kingsolver's homesteading *Animal, Vegetable, Miracle*) justify prioritizing satisfying, unalienated feeding labor and argue that they are also doing their part to fix the food system by telling the story of their own domestic explorations to a larger public. Focusing the lens of do-it-yourself nutrition activism on an explicit moral politics of care around feeding, these authors, like the gardeners I have worked with, argue that small-scale domestic practices like food procurement and gardening—and particularly practices done in social milieux, like gardening with others—provide valuable venues for organizing political action across a range of scales. Contesting received wisdoms of nutrition and dominant modes of food delivery and procurement, these authors harness moral indignation about feeding politics to radical gardening and homemaking practices. They link these practices to a politics of withdrawal from market commodification of feeding. Dedicating themselves in a politically engaged and intentional way to food work that cares not just about the people to be

# Figure 3.1

The back cover of Shannon Hayes' book *Radical Homemakers* juxtaposes feeding and gardening tasks with political economic aspirations: "weed garden" and "drain lifeblood from multinational corporations" are what remain on the to-do list

fed but *all* people and parts of the food system, these radical homemakers suggest that feeding work can be used to address food problems by people who enjoy this kind of work and who are bothered by food system problems. Politicizing domestic work by bringing it public in this way also addresses frustration about individuals' limitations in fixing large scale food system problems—limitations that are due in part to difficulties prioritizing and carrying out food-related work in a way that efficaciously addresses salient problems, given the globalized and uneven power relations of the contemporary food system.

## Understanding good food work through garden stories

The performance of a moral politics of withdrawal from mainstream economies to support alternatives through both practice and narration has become something of a lightning rod in the fraught discursive terrain of food politics. Critics of the claims made for civic agriculture argue that forms of food activism like gardening may be more likely to turn *away* from public politics than to engage with it. Pointing to class, race, and gender inequalities reproduced in some radical homemaking discourse, critics also call into question whether reappropriating domestic practices actually achieves radical outcomes. In their 2011 *Radical History Review* issue on "Radical Foodways," Dan Bender and Jeffrey Pilcher call food scholars to move beyond "comfortable declarations of radical potential." Attempting to respond to this challenge, I am writing this chapter to attempt to make more legible what happens in gardens as people explore and try to reconfigure some of the qualities and social relations involved in the way people access food and the way that the food system is structured. Good food work often combines politics with qualities of pleasurable hobbies. Straddling pleasure and usefulness has benefits: politics and pleasure contribute to each other, making politics more sympathetic and sustainable and politicizing activities people enjoy in useful ways. However, good food work has the potential to become merely a gratifying hobby that is justified by humanitarian motives—and both these hobbies and motives may end up reinforcing white, liberal, middle-class foodie, gardener, or nutritional aesthetics without actually achieving sought-after food system transformations (DeLind 1999, Pudup 2008, DuPuis and Goodman 2005, Guthman 2008). For example, in the way that radical homemaking interventions are oriented toward individuals' responsibility for nutrition and food self-sufficiency and overwhelmingly focused on action in the family home, they tend to reproduce white middle class privilege—even as they struggle with the implications of this privilege (Friedman and Calixte 2009).

I identify this struggle because I think that making this tension explicit is productive—and because I want to help those who engage in food improvement conversations to continue these conversations beyond the defensive silences that can follow from critiques of well-intentioned effort; it seems like laying out clearly what is at stake in these tensions is one way to help. There are no easy answers to the questions of care and responsibility raised by those critical of existing social relations in the food system. Keeping such tensions in mind as we engage in food work (and in talking about food work), however, may build capacity to support systemic care and responsibility around food. So, recognizing the value of both radical homemaking testimonials (and the many ways in which they *do* grapple with their limitations) as well as the critiques of their limitations, I note a few central points that are often raised in relation to radical homemaking and similar work that emphasizes modes of withdrawal from formal economies. First, radical homemakers are seen to reduce emphasis on the possibilities of collective action and regulatory responsibility. Second, their livelihood strategies are usually predicated on social networks that are inaccessible to most people (both because of social closure and also because most people don't have the personal safety nets that allow them to experiment with alternative livelihoods and provisioning methods). And third, they often skirt—or echo without rigorously engaging— significant concerns about problematic devaluations of women's work, especially marginalized women's work, for example by celebrating (e.g. on blogs) the ability to reclaim food work tasks like gardening, cooking, or baking to feed their own children without recognizing the persistent struggles of economically and racially marginalized women around the same tasks—that is, struggles in carrying out such tasks for their own children while they care for others' children.

As marketization displaces most aspects of feeding work onto lowest cost providers, feeding tasks that were once considered necessary, such as gardening, cooking, and food storage, are refigured as optional hobbies that may be virtuous and demonstrate competence but that are also problematically time and input demanding. Gardeners described how this tension is underlined by increasingly visible contradictions in market relations around food, especially as these are called out in popular food politics: the ability to participate successfully in the market relations that ensure access to high-quality, privatized care services such as health care, child care, and education is often predicated on buying into a feeding market (particularly via prepared foods) that many fear is nutritionally and culturally (as well as politically, ethically, and meaningfully) bankrupt. Noting, as Bender and Pilcher (2011) also do, that food work is often dismissed for being "too fun," many of the people I interviewed while working on garden projects were explicit in contrasting romantic, feel-good, consumerist desires for short-supply-chain aesthetics with practices that not only keep food skills alive but also maintain critical analyses and build social relations focused around ideological and material supply chain challenges. Many gardeners were critical about the limits of their own actions, fretting that the diffuse practices of food politics often fail to change the problems at which they are aimed. This concern and the ways gardeners describe and act on it (which I describe below), suggest potentially empowering ways that gardening allows people to explore their relation with the food system, and to become more aware of the ambiguities and politics of food work than they may get credit for—even if their politics remains centered on gardens. Especially in the context of public garden projects, the discourses in which such concerns circulate highlight how gardeners explore many aspects of social organization that may seem subsidiary to food production, but that may, in fact, be crucial to understanding the social context of nutrition. In trying to assess whether garden activities move beyond being merely a satisfying hobby to meet larger goals, it seems worth recognizing the long history of public gardens as projects meant to demonstrate the provenance of food—as well as the care that should be taken to grow food properly;[2] this function of garden work responds well to anxious concerns about the alienation of eaters from knowledge and power in the contemporary food system, and suggests some of the less obvious ways gardens might be assessed as useful. The discourse and practice of food and nutrition, situated within particular social contexts, certainly shape garden and other food projects. Careful attention to the way that people participate in gardening discourses may enable those interested in critical approaches to nutrition to better engage with public garden projects, and to foster collaborations that may meet critical nutrition goals as part of such projects. Below I pay careful attention to the way that people reproduce garden stories—stories that may contain and convey critical organizational tactics and strategy, but may also carry considerable baggage in terms of assumptions about hierarchical social control.

Along with preparing food and procuring food through shopping (or trade), food gardening practices may put otherwise non-agriculturalists in close proximity to the way in which the majority of the world has historically been fed: through feeding labor done predominantly by women, often in close association with household labor, and often in what is now deemed the "informal" sector (Butterfield 2009, FAO 2011, Fortmann 1980, Marsh 1998). For some gardeners, this connection to women's labor adds a layer of solidarity and political meaning to their gardening practices that may encourage them to connect their gardening with more systemic concerns and political practices. For others, however, gardening is a means to a self-sufficiency that is asserted as *apolitical*—a form of withdrawal from the problematic social, political, and environmental contexts of contemporary food and nutrition. I have followed the varied stories that people tell about why and how they garden in public contexts in an attempt to better understand this tension between public and private—between taking on food politics in the confines of the garden, and taking the politics of feeding *public*. Garden stories show us how the politics of feeding moves from the home spaces where food is imagined to be properly procured, prepared, and eaten, through a series of social networks that form around gardening *outside* the conventionally imagined backyard-to-kitchen produce supply chain.

I examine these garden stories against a backdrop of the rise of "radical homemaking," a practice of self-sufficient homemaking that attempts to more explicitly account for many of the social, political, and environmental values

---

2   It is worth noting that growing food properly is not only framed in terms of progressive good food politics but that nourishing the family and even the larger society (often the nation) has also been used in the service of more complicated politics, particularly around food work for war and around contentious race and land politics.

involved in everyday provisioning within household ecologies. (I consider radical homemaking as a contrast with public garden projects rather than as my central focus; see also Deutsch 2011.) The failure to account for social, political and environmental values has been usefully portrayed as problematic externalizations of cost not reflected in the price of food in the contemporary modern capitalist food system. Such externalities include soil degradation and chemical toxicity, health effects of pesticides, poor diet, poverty, and unjust labor practices; gardeners have described internalizing these costs by normalizing soil and ecological stewardship, healthy food and foodland access, and food production and governance as part of people's everyday experience of food. However, following Michael Mikulak's work, I hesitate to limit the values in question to econometrically measurable costs, or to suggest that values not central to food market mechanisms should be commodified, especially since so much of what is radical about food improvement projects involves the decommodification of food values. These values resonate with food-reform oriented critiques of contemporary food and nutrition that have also been framed in terms of a "good food gap," defined eloquently by Lauren Baker as "the policy space that exists between the farm income crisis and the health crisis," in which "farmers find it hard to make a living growing food and consumers find it hard to make the good food choices they want to make" (2010). I suggest that it is helpful to pay attention to *how* and *why* the stories that are told about practices of gardening *matter*—in both public and private domains—to rethinking nutrition in terms of the way that people both experience and talk about food and feeding. It is important to recognize that people engaging in gardening and feeding often have a view of health that extends beyond their own bodies to include health of landscapes and communities. In this understanding of feeding work, nutrition becomes much broader than simply an intervention focused on individual eating, and it is more directly connected to politics and social justice.

## Gardens, nutrition, and food politics

Food gardening is a significant focal point of current movements to change food systems. In addition to providing fresh food for consumption and storage, gardening for food is considered to be a good way to support and expand people's capacity to understand and care about food. Further, food gardening is popular because it provides opportunities to integrate a number of different approaches to improving nutrition: especially when done in public collaborations, gardens offer a venue for exploring and propagating methods for improving a wider range of food system parts than may be obvious—from access to fresh produce to food system governance models. Such food systems improvements are the basis of what people are calling "the food movement" or "the good food movement" (Baker 2010).

Food gardens have historically provided foundations for such politics around improving food, and have recently been returning to popular awareness in this role. The politics organized under the broad and integrative umbrella that gardens provide tend to share three interrelated and salient features that I introduce here and then expand upon briefly below: (a.) the garden work is justified at least in part by claims about nutrition; (b.) women do much of the work of gardening and subsequent food preparation; and (c.) the gardens are important not only because of their material existence but also through the stories that are told about them. In this chapter, I focus on these stories that are associated with the social organizing of food gardening in order to explore the aspects of food gardening that may be useful to consider in nutrition studies. I argue that gardening is of value not only for the obvious reasons of improvements in food security and access to fresh foods (Alaimo et al. 2008, Marsh 1998), but also for the ways in which the mundane performance of gardening provides a focal point for participating in social dialogue around problems of food, feeding, and nutrition. Reflecting on the dramatic resurgence of stories about gardening over the past decade, I examine the function of garden stories as a mode of food politics that provide gardeners and their allies a venue for exploring, organizing, and forming solidarities around systemic food improvements, not just better food for themselves and their families.

## a. Nutritional gardening

Although they may significantly impact nutrition, many of the most important problems in contemporary food systems are not directly problems of nutrition. For example, although malnutrition, poisoning, or metabolic problems might result from the exploitation of workers or the pollution of soil, air, and water by agricultural chemicals and byproducts, labor arrangements and pesticide use are not problems of nutrition per se (Guthman 2012, Perfecto 1992). However, despite considerable educational and popularizing work over decades focused on improving systemic problems with food, people often appear more motivated to address food issues when they are tied to nutrition—and to taste, in both the senses of sensory qualities and of signaling social distinction. Arcane aspects of food science, particularly having to do with the extension of shelf life, for example via preservatives or hydrogenation, or the substitution of widely available agri-industrial food inputs, such as corn or soy derivatives, have moved from laboratory vocabulary into the battle cries of food activism, and have served to give people outside the food industry glimpses into the complex relationships that make up the food system, and the ways that these are changing what we ingest (Schurman and Munro 2010, Guthman 2012). As global agri-food regimes falter in providing healthy, affordable food and farmer livelihoods at the same time that journalistic and academic writing about food are increasing popular, everyday food encounters like gardening become sites to explore concerns about recurring food crises. Hence, while gardens may be associated with good nutrition in the most narrow sense, they also help us to explore how problems of nutrition are situated within social and ecological systems that gardening may affect and/or be affected by.

## b. Gendered gardening

As people learn about the food system, the role of *gender* is something many of the people I've interviewed (as well as most of my students) find striking. Despite the desire for gender equality that is often explicitly expressed in food system work, and despite significant anomalies in the general pattern, most of the people I work with acknowledge the persistence of stereotypical gendered division of food labor: they describe a widely shared image of women as responsible for food within the household (even if they express optimism that this may be changing) and of farmers as aging white males. Thus, people who have encountered evidence that women are primary food producers in many cultures tend to underline this as significant, and draw exploratory parallels between global trends and their experiences of feeding work such as food gardening. Food gardening has traditionally been a significantly gendered activity, often considered women's work even within the specifically Anglo-American contexts where I conduct research (and where men also have gendered roles in producing yards and gardens). As feminist scholars in a wide range of different fields having to do with food and feeding have pointed out, women tend to be disproportionately responsible for the work involved in obtaining nutrition (DeVault 1991), what I refer to as *feeding work*. This pattern of overrepresentation of women in food procurement extends not only to food gardening, but also to contemporary food activism more broadly, and by "food activism," I mean the myriad ways that people try to figure out how to improve some of the problems with contemporary food systems sometimes described as "externalities" (DeLind 1999).

## c. Storied gardening

As the central focus of this chapter, "storied gardening" refers to people's *descriptions* of how they use home, public, and institutional food gardens to address social and ecological problems related to food. I consider the relationship between the kinds of gardening practices that people use to critique or improve food systems and the stories that they tell about this food system reform through their experiences with *other* people's gardens. I focus on the relations between stories and gardens in part to acknowledge and explore the difficulty of separating the material food production practices that take place in gardens from the more narrative and discursive functions that food gardening enables. I analyze the relationship between practices and discourses through my encounter with a series of gardens that have become entangled with food system reform efforts, and with the related stories that are told and reproduced to effect food system reforms.

Two key goals represent much progressive food reform effort: food sovereignty (i.e. control over the food system) and shorter and more transparent supply chains. These appeal to people because they promise to address many of the problems associated with externalized costs of food. However, even amongst those who are interested in and aware of complex issues of food and nutrition, most struggle to identify what leverage points their actions could possibly affect. Consequently, a significant function for much food activism has a heuristic nature, rather than a primarily functional one. I focus on *other* people's gardens, rather than the garden-variety backyard garden, because efforts to work with others to realize the potentials that gardens represent illustrate the nature of the relationship between the praxis of small everyday activities in relation to complex global food regimes. The extension beyond one's own everyday space into collaboration with others is easier to examine and to denaturalize in order to analyze—and the challenge of trying to figure out what gardens *can* do, and what they *are* doing (perhaps by considering the subsidiary or "off-label" benefits of gardening) demonstrates the open and uncertain nature of the relationship between garden stories and the externalities that people are trying to deal with.

The degree to which gardening has become a site for taking on the challenges of global agri-food regimes may come as a surprise to many—where in food gardens do we find solutions to larger food system problems? Many critics suggest that gardening, like many other approaches to improving food systems, encourages food reformers to focus too much on themselves and their own food needs and not enough on the broader structural and social arrangements, such as racism and capitalism, that maintain injustices in the food system (Pudup 2008, Guthman 2008). If these problems have so much to do with the structural arrangements of large-scale food production, as many argue, doesn't food gardening encourage gardeners to imagine themselves sidestepping responsibility for these problematic arrangements (and mostly symbolically, at that, since proportionally so few gardeners procure much of their nutrition from their gardens)?

Those who defend the value of food gardening as a mode of addressing structural food problems tend to argue in favor of gardening largely by asserting the value of food sovereignty and of reducing the length (and improving the quality) of the supply chains between producers and consumers. The "short(er) supply chain" argument is based on the idea that reducing the number of steps between producers and consumers helps create social relations that internalize the costs of food production, particularly those associated with food processing,such as unfair labor conditions, the introduction of unhealthy additives to food, and the profligate use of energy to reshape foods into more storable, "value-added" forms. Proponents of short supply chains also argue that minimizing the distance "from farm to fork" helps sidestep many of the problems that arise along food chains, which, by this logic, are enabled by their invisibility amidst the opaque complexity of global agri-food regimes. The food sovereignty argument even more directly addresses the question of control over food, suggesting that the right to food is best assured by direct political relationships between producers and consumers, focusing primarily on political arrangements that enable people to remain (or become) food self-sufficient. While critics of these claims point to the difficulty of validating how effectively gardening contributes to such goals (is corporate agribusiness changed by gardening?), I argue that the larger systemic motives attached to gardening are important at least in part because they allow us to understand the stories that people tell about their gardening as interpretive acts. Bringing into focus the discourses that frame political gardening may contribute to the efficacy of stories in connecting with salient audiences, both by making the interpretive frames garden story tellers use more legible to people in a range of different positions in the food system and also by supporting space within garden narratives for critical reflection, particularly around narratives that are likely to express (or to be interpreted in terms of) conservative defensiveness, self-righteousness, smugness, or self-centered political withdrawal, all charges that have been leveled against neo-agrarian sentiments (Freyfogle 2001, Guthman 2008, 2012, DuPuis and Goodman 2005). In the next section I trace critical themes in neo-agrarian garden stories, exploring the question of what garden stories appear to do, and why what they do is important to "doing nutrition differently."

## Proliferating garden stories

I came to focus on gardening in my research because it tends to be the land use most often brought to my attention when I talk to people about their interactions with landscapes. The stories I am told (regardless of what I ask or do not ask) have made increasingly clear to me how important talking about gardening—and performing what is talked about—can be to the significance of gardening. Gardens are not merely illustrations of stories, nor are they just demonstrations of food system improvements that primarily produce a different set of food and nutrition outcomes. Instead, they host a complex interplay of politics and ecologies that support the exploration and propagation of a range of food practices. Functioning in a more subtle register than direct affronts to the mainstream food system (such as the regulation of transfats or attempts to change state nutrition guidelines or enforcement of labor laws), everyday practices of food production may help to decenter and enrich both popular and academic understanding of how people feed themselves and others—as well as how they imagine themselves doing so, an imagination that may reveal as much in its aspirations as it does in its shortcomings.

In this section, I examine cases via the frameworks of "critical neo-agrarianism" and "other women's gardens." What I've previously called "critical neo-agrarianism" frames a specific critical strand of the current interest in agrarian issues (what's been labeled "the new agrarianism"), largely through addressing food issues in their social contexts and focusing on building solidarity and collective action networks to effect food justice. I start with the observation that people who engage in garden politics, and also those who do not, both appear to underestimate the substantive work that garden stories do. I suspect that this has much to do with difficulties in translating not only between the efforts people put into gardening and the political claims they might associate with those efforts, but also between the hopes they may have for gardening and what seems possible. The literature on "radical homemaking" provides the most dramatic claims for what gardening might be able to achieve—and not unrelated to the big ambitions that are explored in this literature, it is also subject to considerable ire and critique. In the context of charges that the politics of self-sufficiency espoused in this literature represent problematic oversimplifications of the challenges faced by those marginalized in contemporary food systems, it may be useful to recognize the dramatic ways that radical homemaking and back-to-the-land narratives catch hold of popular imagination and circulate in popular knowledge. In part, this may be because of the way that they capture white middle-class rural idyllic escape fantasies and sympathies more than food justice narratives do.

I have come to use the phrase "*other women's gardens*" to represent my line of research on the ideologies, practices, and narratives of gardens and their use in on building solidarity and collective action to approach food improvement efforts. This phrase was inspired by the Lord Wavell book *Other Men's Flowers* (1945), a copy of which was gifted to me in the early stages of my work in Aotearoa New Zealand by Rhonda. Her favorite book, *Other Men's Flowers* is a collection of poems that Lord Wavell selected, memorized, and anthologized during the Second World War. As I have struggled with the fact that most accounts asserting the importance of garden and many other food system interventions rely heavily on anecdotal stories that seem insignificant or even self-servingly escapist or apologist in the face of overwhelming quantitative data about the problems to be addressed, *Other Men's Flowers* has become an important metaphor for the way that some men and an overwhelming number of women have demonstrated their garden stories to be creating significant spaces of political praxis by extending their garden aspirations through each other's gardens and daily lived practices. I imagine Wavell prioritizing the practices of selecting, memorizing, and publishing poems of wonder and hope during a time of war not in order to withdraw from the political economy of the situation, but to seek, explore, and proffer potential ways to remain present in the world at a challenging time of despair and violent upset. Further, in the context of good food work, which is foundationally situated within the need for food justice, I find the contrast between an extraordinarily privileged white man's wartime poetry collection and the focus of critical race theory on storytelling to be a productive tension. Stories of garden politics sometimes sound problematically conservative in part because they are constructed and circulated in social contexts that suffer from the problems of conservatism.[3]

---

3  Gardens can be powerful reminders of the colonial histories they manifest in their many layers: in the layouts, foods, justifications, rules, social relations, and aspirations encoded there. They are also sites of resistance to continued colonization, in terms of both the eco- and social- aspects of agroecology—and the right to food associated with the food sovereignty aspects of the agroecology tradition. The relationship between white charity models and critical neo-agrarian models remains a significant tension in public food garden projects. Gardens provide such a significant site for talking through some of these issues partly because they are so filled with them and also because the co-existence of conservative and progressive motives for and modes of gardening (and the sympathetic space afforded

Both withdrawal into the comforts of home, on one hand, and engagement with uncomfortable realities through the rehearsed performance of varied food and feeding practices, on the other, take place in the modern garden. For the people who have shared their stories with me, garden stories are an important venue for representing exploration and learning in the face of this tension between withdrawal and engagement. Such exploration, they assert, leads to the sharing of food and of feeding values and practices, and also to the creation, maintenance, critique, and validation of alternatives. As I have struggled to understand how the telling of stories of change effects political or material changes to food systems, *Other Men's Flowers* has morphed for me into the idea of *other women's gardens* as I have listened to thousands of stories about gardens and thought through why the person-to-person sharing of anecdotes of food system improvement through the garden has played such a central role in the ways that women, particularly, have asserted their agency in the food system through the relational networks that situate household labor in broader systemic context. In the following three sub-sections, I explore the moral politics of feeding produced in garden work through a brief analysis of some key stories about the effects of gardens, in the context of the emerging literature on "radical homemaking," paying particular attention to the critiques leveled at garden politics by those who focus on the limitations and disengagements involved in gardening as a form of politics—as well as by others trying to study and practice feeding work differently.

**Listening to garden stories: Exploring "critical neo-agrarianism"**

I have spent much of the last decade reading and listening to stories about other people's gardens, as part of ethnographic and survey projects about food systems and land use conducted in several different sites: in and around Toronto in Ontario, Canada (50 households surveyed over 6 years: Cadieux 2001, 2005); Minnesota (50 households over four years) and New England (20 over two years) in the U.S. (Cadieux 2007); and the Canterbury region near the city of Christchurch in Aotearoa New Zealand (50 households over 1 year: Cadieux 2006, 2008, 2011). The food and garden projects I have researched and volunteered with fall along the full range of rural to urban locations, and in all sites include both household as well as institutional food and garden projects. These cases exhibit most of the features of contemporary urban and suburban settlement form and also food system activism, which I encountered by talking with people across a wide range of landscapes being gardened (in addition to surveying other parts of the food system outside the scope of this chapter). Although contrasts could be drawn between them, especially about engagement with food's global and systemic contexts, these cases exhibited many similarities across the board as well, especially around

by shared interest and shared practices in gardening) support potentially reflexive discourse that can be mobilized to help people think across scales in food system politics.

the politicization of food and the exploration and performance of food politics through gardening and the propagation of garden stories.

The gardens I surveyed were often home gardens, although most of the gardeners who spoke with me were also involved in food system work beyond their own domestic spaces, many through more socially-oriented gardening—and I focus here on those more public gardens, what Mary Beth Pudup calls "organized garden projects" (2008) (e.g., gardens associated with community centers, food access organizations, churches, schools, and neighborhoods, including both public "community gardens" and private land-sharing arrangements or associations between neighbors, of a range of formalities). My encounter with gardeners has been mostly situated in the context of ethnographic research or social engagements of at least several months, if not years, in which I lived in or near the places in question, worked in the gardens, interviewed people I worked with, and also sought out other people mentioned as good sources of knowledge about gardening, urban land uses, and, as gardening has increasingly taken on the contemporary language of "food systems," about popular involvement in food work.

My explicit interest has been in the persistence of land-use strategies that are neither entirely justified for either pleasurable aesthetic or practical economic purposes, but that rather involve a blend of engaging and appreciating the combined productive and amenity features of land uses via "hobby" farms and urban agriculture. Such deliberately performed land uses are often somewhat exceptional or out of place in landscapes where they are found, landscapes that tend to be comprised of urban lawns or rural agriculture. Partly because people use these slightly more controversial landscapes to explore and call into question dominant environmental management strategies having to do with urban and rural land use planning, these landscapes also tend to be politicized. I am particularly interested in the tension that exists between the idea of using everyday environmental interactions explicitly as politics, on one hand, and the charge that is often leveled against people making progressive claims about their environmental politics, on the other: that land uses like hobby farming do *not* represent progressive politics, at least not effectively so, and are rather a *withdrawal* from politics into a hard-working but fundamentally apolitical or conservative lifestyle (Cadieux 2005, 2007). My research set out to try to differentiate *critical* strands of the new agrarianism from celebrations of the resurgence of agrarianism that did not address problematically exclusionary aspects of agrarian culture. And so it was through researching this tension that I came to hear so many stories about the politics—and particularly food politics—involved in particular land-use choices (making food systems more locally controlled, transparent, ecologically healthy, and just), and that I was drawn in to the moral politics directed at food systems from gardens.

All of the sites I researched have strong, diverse, and politicized histories of food production across a range of scales, including conspicuous emphasis on home gardens, both in historical and contemporary context (Warner 1987, TCGN 2012, Ristau 2011). At the apex of Canada's highly productive and southern stretching "banana belt," Toronto is often described as being situated on one third of Canada's best-ranked agricultural soil, and widespread urban agriculture takes good advantage of this resource (Lister 2007). Both urban and rural areas of the Upper Midwest and New England in the U.S. have long and currently very active traditions of food gardening, across a range of different class, ethnic, and gender positions, with gardening only sometimes associated with food movement critiques of the mainstream food problems so often highlighted in current press on the new agrarianism. Many of the touted gardens in question were flaunted as much for their *frontyard* ornamental prowess than their backyard practicality (with significant accompanying class and colonial connotations to garden choices). Christchurch has traditionally been known as the "Garden City" of Aotearoa New Zealand, with a dominant urban form of quarter acre lots and a rich tradition of urban and suburban gardens—a garden researcher's dream. Although I could use reports from any of these sites, I will focus on some examples from Christchurch to illustrate the key themes of the garden stories I heard, and then to consider the implications of public garden projects.[4]

As I have documented elsewhere, what was most striking about the Christchurch gardeners who spoke with me was their situating of their everyday garden practices in systemic contexts. From soil systems to global political economy, gardeners narrating their practices took pains to make sure I understood that their gardening needed to be understood in broader historical, geographical, and political contexts (Cadieux 2006, 2008). Their organized garden projects have missions that often explicitly include addressing externalized costs of mainstream food production: many trace their lineage to Soil Association roots, with associated concern about ecological health; some highlight making healthy food and food production practices more accessible to people whose access to food or foodland might be marginalized; others aim to address ecological and cultural damages of colonization through regenerative gardening practices by incorporating indigenous plants and food traditions.

Individuals frequently framed the broader implications of their gardening practices with reference to their participation in or knowledge about such garden projects. They pointed out that while their small garden patches might not amount to much, they provided a way to participate in the food system by, at the least, giving them a domain to materially experiment with and better understand the implications of how food might be produced. Even more ambitiously, gardeners described reminding themselves through their everyday gardening behaviors of the topics and practices on which they ought cooperate (or at least have solidarity)

---

4   "Garden" in the Christchurch context tends to mean the whole yard, hence the focus on ornamentals; although I focus on food gardening, I do not necessarily draw a strict boundary. The earthquakes around Christchurch over the past two years have changed the gardens and land uses I describe dramatically; however, some commentators have suggested that the social networks involved in environmental and food politics that I describe have played a role in figuring out how to deal with the earthquakes' continuing effects, both in terms of short term response and in developing longer-term eco-social resilience.

with others in order to effect change. Even if they were modest in assumptions that their own practices were a sure pathway to sustainability and generous in their assessments of mainstream commercial food production—sometimes going so far as to assert that the ecological and social damages associated with large-scale agriculture were beyond the ability of farmers to internalize while still producing the food expected of them—almost all of the gardeners who spoke with me interpreted at least some of their gardening practices as explicit food system critiques. Three types of gardening aspiration from the Christchurch case illustrate ways that food reformers attempt through their gardening, and narration of it, to define the politics of feeding against what many gardeners explicitly named as an oppressive and hegemonic mainstream food system.

First, gardeners called property into question in a range of ways, from demolishing the fences or hedges that commonly separate Christchurch residential properties from each other and joining gardens into less bounded common space to organizing work crews to care for others' gardens (sometimes surreptitiously, for the benefit of elders struggling to maintain gardens). People often took me across boundaries while they talked to me, showing me where boundaries marking different modes of garden management were maintained or where they had been effaced or made more (or less) permeable, as well as where native and imported plants were and were not mixed (and why), and showing how their social associations around ecological restoration and food gardening were helping to contest the assumptions involved in associating garden practices with spatial and social segregation. I begin with this example because it exemplifies the conflicted nature of much garden activism: although dramatic examples—of whole blocks joining backyards into common gardens or regenerated forests or of social justice organizations gardening common properties to make food freely available to those in need—were narrated enthusiastically by the majority of people who spoke with me, only a small minority actively participated in them. This illustrates the way that certain aspects of garden aspirations often remain unmaterialized; their narratives function as aesthetic inspirations, and this inspiration helps to motivate material action, but the alignment between progressive ideological intent and the outcomes of gardening practices is often shaky. Where the stories seemed to help effect significant changes were in the unusual cases where the stories encoded instructions that helped resolve uncertainty over what to do in various practices of gardening, such as when an adjacent garden of an elderly neighbor needed care in a neighborhood with active garden sharing; most garden tasks, however, are not particularly uncertain.

Second, gardeners called the colonial nature of the food system into question by attempting to incorporate decolonizing practices into garden management. Although often (but not always) recognizing that their food was implicated in both consumption and production (and hence geopolitical) relations with global food regimes, gardeners emphasized the way that they could choose food plants, organize the cultures of their gardens and provisioning, and publicly interpret their gardens in ways that decolonized food flora, landscapes, and social relations. Most of the common-property gardeners used their gardens as platforms to demonstrate alternatives to land uses or food system practices they interpreted as reproducing problematic class, race, and gender differences, for example the reinforcement and reification of separations between settler society and indigenous by valorizing large-scale food import and export, or by filling garden landscapes with imported food plants and flowers, or justifying separations between classes by defining as "proper" status-granting plants like roses and lawns (also imported)—or between genders by assuming different spheres of responsibility in the garden and kitchen.

Again recognizing that their contesting of land use and food system hegemonies were partial and that they were often complicit, for example, in supporting food trade or in Maori dispossession, these gardeners engaged in considerable educational outreach about their garden practices. Through signage, brochures and websites, collaborative work days, and organizing volunteers for both informal and also municipal- and regional-government sponsored planting days, the most radical gardeners emphasized the social relations that undergirded their plant and land tenure choices, and drew attention to the way that the organization of their garden landscapes (via collaborative governance, for example) helped reinforce solidarity against reactionary defensiveness around colonial norms and tastes defining "the Garden City" landscape aesthetic.

These were also the gardeners investigating cultivation and social practices that would enable more use of native edibles. In contrast to merely fetishizing locality and indigeneity, however, in seeking rapprochement with local tribal groups who held jurisdiction over use of native plants, gardeners brought to the foreground the cultural and ecological regenerative potential of gardening as an inviting and potentially non-threatening step toward understanding responsibility for the disruptions of colonization. In a domain marked by frequent distancing from government regulation and intervention (part of my concern with investigating a critical version of the new agrarianism had to do with the anti-egalitarian libertarianism of much agrarian rhetoric), this group was most likely to invoke the potential value of state and local government intervention in mediating contests between different interests. For example, they described city and federal values-based planning paradigms, and suggested that their interest in these approaches to planning—and willingness to grant them legitimacy—was perhaps due to what they had learned through intensive grappling with competing (and often incommensurate) land use, resource, and relational philosophies at a range of different scales.

Third, while those gardeners addressing colonization were exceptional, *most* gardeners made a point of explicitly connecting their gardening to larger political contexts and goals. The kinds of stories people seemed likely to tell about their gardens could easily be evaluated as being more aspirational than most narratives describing other similar types of socio-environmental aspirations. (This was demonstrated, for example, by the lofty goals of deconstructing private property and colonial social organization versus more prosaic everyday goals of increasing waste recycling or reducing water use; people did not tend to wax as eloquent about most waste and water programs I encountered as they did about gardens.) Gardeners went out of their way, however, to demonstrate beyond "mere talk" the way their gardening practices gave them a way not only to mitigate food system impacts but also to scale up, politically, to instigate systemic change. More politically oriented gardeners described the way that garden practices—such as composting to recycle nutrients and organizing to ensure access to food and foodland—gave them concrete ways to address food-related social issues in common with others in ways that contributed to political organizing, from neighborhood to party (particularly Green Party) to international scale (including efforts to change WTO rules to promote more equity for the Global South and to apply pressure on the U.S. via the UN and NGOs to abide by rules governing labor and chemical use for food production). And even gardeners who were not particularly politically active (and further, who represented a wide range of political stances) made a point of making sure I understood the larger symbolic but also material political contexts and goals of their gardening (e.g., the way that their everyday food production was situated in the broader political and economic contexts of the country's position in global food trade). They also emphasized the ways that the social valorization of food production, the institutionalization of home gardens via the quarter acre urban lot, and the popular enthusiasm for environmental sustainability and outdoorsy aesthetics had to do with British colonization, shifting global geopolitics, the neoliberal turn away from farming subsidies, and the uncertain relationship between state, civil society, and international trade that made people attentive to ensuring diverse strategies for procuring food, both at the level of the household and the national economy.[5]

Gardeners repeatedly emphasized the way their practices created space (both in their own experiences and in their social interactions) to question the assumptions that make the inputs and outputs involved in everyday feeding invisible. They described their material attention to soil fertility and human health as connecting vegetable peelings and mundane agroecological decisions (e.g. how to deal with competition, water, plant choice, or prioritization of different peoples' goals and tasks) to a systemic perspective that helped counter the simplified commodification of food that they saw contributing to the food system making people and ecological systems less healthy. The links between this common description and the practices that were used to carry it out were demonstrated repeatedly during my time with gardeners. They paused our conversations to go add kitchen scraps (or urine) to compost piles or pointed out neighbors carrying kitchen scraps to others' gardens. Or they pointedly expressed the ways their nutrient management not only

---

5    Some of the most common starting places, across the spectrum of interviews I conducted, were represented by description such as: "you realize, the British only arrived here about a hundred and fifty years ago;" "and when Britain joined the EEC in 1972, everything changed, because we were on our own, having to fend for ourselves in the global market;" "and everything changed again in 1984 with the 4th Labour government and the turn toward neoliberalism."

stewarded resources but also gave them a platform to act out for themselves and others the implications of resisting comforts and conveniences they felt contributed to distributing resources in illegitimately disproportionate ways. All of these examples illustrate the ways that gardening practices enabled people to assess and attempt to account for the costs associated with their own material existence, and the way they *described* internalizing the food system impacts that might otherwise be externalized shows how they leveraged their garden experiences to try to push for a broader adoption of such internalization, and of recognizing broader values around food.

I was impressed by the thoughtful reflection that many gardeners provided on their gardens as domains for trying out, exploring, and practicing ways of struggling against oppressive aspects of class relations, settler society heritage, and global and local manifestations of neoliberal capitalism and accumulation strategies. I was also impressed by their acknowledgement of the limitations of garden practices for effecting systemic change, even if such practices created and maintained spaces of resistance that were explicitly social and political in the ways they linked individuals, institutions, and "the food system." As an academic, and as someone who came to gardening at least in part through written stories (and through the written stories that convinced my friends and parents to garden), I was also sympathetic with the dilemma gardeners encountered in the way they represented their gardening: on one hand, they felt compelled to document and narrate their gardening—many had newsletters, blogs, pamphlets, or signs describing their gardening rationale, and even less organized garden projects had often-told stories that were frequently repeated to me by people with different relations to the gardens and gardeners. On the other hand, in reflexive moments, gardeners expressed reservations about whether their narration of aspirational motives for gardening was too optimistic, giving desired audiences a sense of false accomplishment and comforting self-satisfaction, rather than serving to support commitment to ongoing practices in service of addressing challenging food system problems.

## Concluding thoughts

In the process of developing the analysis I present here, I have come to find considerably more solidarity with what many gardeners are doing to improve food systems. Even if the material and discursive practices of gardening are not always mobilized in progressive ways, many gardeners are using gardens to explore potential empowerment in relation to food. Many garden stories may be more socially radical than they are credited for being. People bring these stories into public domains because they wish to highlight some aspect of gardening practice that seems useful to others, and radical gardening stories may be useful for addressing food, feeding, and nutrition in a number of different ways. For example, stories about acts of engaging with other people in order to garden have very different qualities than stories that celebrate the heroic self-sufficiency of home gardens. Gardening with others changes what happens in the garden: ecological or social challenges that might be more likely to be ignored or avoided in home gardens are more likely to have to be addressed when gardening with others, and the intentions and practices that link everyday gardening to larger contexts are more likely to be witnessed and to enter into dialogue with others' intentions and practices. This dialogue generates a kind of eco-social learning that builds adaptive capacity and resilience in part because it helps people apply some of the systemic lessons learned in dealing with natural challenges to social challenges, and vice versa.

Thus, part of my closing argument has to do with academic focus; focusing on the garden work that makes change—and on the many things produced in gardens beyond produce—may help gardeners tell stories about actions that build different capacities for good food. Food justice stories, particularly, help to highlight another implication of publicly performing food politics: talking about gardening in terms of who gets to speak, who gets to tell what stories, and about who should hear the stories (as well as to what actions and histories these stories should be tied) has the powerful capacity to make more space for justice in food discourses. Stories circulate knowledge and power in ways that can both reframe the food system (away from the silencing and traumatic configuration of the contemporary dominant food system) and also remind diverse food system actors to listen more actively for others' stories that put food work and politics into social context. I argue that by exploring gardeners' storied practices we can better understand how food work attempts to improve the food system and how it enables a different starting point for thinking about nutrition politics. The performance of gardening and other food work provides opportunities to try out different ways of navigating the social tensions (and contentious politics) of household- and community-scale feeding practices about which critical scholars provide valuable analysis. Gardening, then, becomes much larger than an icon of healthy eating, emerging as a central space for understanding the multiplex and multi-scalar politics of feeding.

## Acknowledgements

Thanks to the participants in and organizers of the How We Talk about Feeding the World workshop and Agri-food reading group, particularly Rachel Slocum, Tracey Deutsch, Rachel Schurman and Jeffrey Pilcher for support during the writing of this manuscript, and even more to Heidi Zimmerman and Jerry Shannon (as well as to Evan Roberts, Sheela Kennedy, Sarah Flood, Natasha Rivers, and Ben al-Haddad) and Allison Hayes-Conroy and Jessica Hayes-Conroy for supportive comments on drafts, to Maria Frank for research help, and the students in Food, Culture, Society for their thoughts on gardening and food politics. Thanks also to Shannon Hayes for permission to reproduce her book's cover. Most of all, thanks to the gardeners whose stories and practices motivated this chapter—I am impressed by the radicalism of your efforts, and I hope this kind of story telling helps work toward the goals you so eloquently describe and embed in your gardens.

## References

Alaimo, K., Packnett, E., Miles, R. A., and Kruger, D. J. 2008. Fruit and vegetable intake among urban community gardeners. *Journal of Nutrition Education and Behavior*, 40(2), 94-101.

Astyk, S. 2009. *A nation of farmers: defeating the food crisis on American soil*, New Society Publishers, Gabriola Island, B.C.

Baker, L. 2010. *10 Good Food Ideas*. Available at: http://www.harthouse.utoronto.ca/culture/tedx-future-food [accessed 11 April 2012].

Basset, T. J. 1981. Reaping on the margins: A century of community gardening in America. *Landscape*, 25, 1-8.

Bender, D., and Pilcher, J. M. 2011. Editors' introduction: Radicalizing the history of food. *Radical History Review*, 2011(110), 1-7.

Butterfield, B. 2009. *Impact of Gardening in America*. National Gardening Association, South Burlington, VT. Available at: http://www.gardenresearch.com/files/2009-Impact-of-Gardening-in-America-White-Paper.pdf [accessed 15 April, 2012].

Cadieux, K. V. 2001. *Imagining Exurbia: Narratives of Land Use in the Residential Countryside*. MA Thesis, Department of Geography and Planning, University of Toronto.

Cadieux, K. V. 2005. Engagement with the land: redemption of the rural residence fantasy? In S. Essex, A. Gilg and R. Yarwood (eds), *Rural Change and Sustainability: Agriculture, the Environment and Communities* (pp. 215-229). Oxfordshire: CABI.

Cadieux, K. V. 2006. Amenity and Productive Relationships with 'Nature' in Exurbia: Engagement and Disengagement with Urban Agriculture and the Residential Forest. Unpublished Ph.D., University of Toronto, Toronto.

Cadieux, K. V. 2007. Beyond the rural idyll: Agrarian problems and promises in exurban sprawl. Yale Agrarian Studies Seminar paper, Yale Institution for Social and Policy Study (16 February) http://www.yale.edu/agrarianstudies/colloqpapers/18conspicuousproduction.pdf.

Cadieux, K. V. 2008. Political ecology of exurban 'lifestyle' landscape at Christchurch's contested urban fence. *Urban Forestry and Urban Greening* 7(3): 183-194.

Cadieux, K. V. 2011. Competing discourses of nature in exurbia. *GeoJournal* 76(4): 341-363.

DeLind, L. B. 1999. Close encounters with a CSA: The reflections of a bruised and somewhat wiser anthropologist. *Agriculture and Human Values*, 16(1), 3-9.

Deutsch, T. 2011. Memories of Mothers in the Kitchen: Local Foods, History, and Women's Work. *Radical History Review*, 110: 167-177.

DeVault, M. L. 1991. *Feeding the Family: The Social Organization of Caring as Gendered Work*. Chicago, University of Chicago Press.

DuPuis, E. M. and D. Goodman. 2005. Should we go "home" to eat? Toward a reflexive politics of localism. *Journal of Rural Studies*, 21(3): 359-371.

Food and Agriculture Organization (FAO). 2000. The state of food insecurity in the world 2000 (section: Dynamics of change: The dividends of food security) http://www.fao.org/docrep/x8200e/x8200e05.htm/ftp://ftp.fao.org/docrep/fao/x8200e/

Food and Agriculture Organization (FAO). 2011. *The State of Food and Agriculture 2010-2011 Women in Agriculture: Closing the gender gap for development*. Rome, FAO. Available at: http://www.fao.org/docrep/013/i2050e/i2050e.pdf [accessed 21 April, 2012].

Fortmann, L. 1980. *Tillers of the Soil and Keepers of the Hearth: A Bibliographic Guide to Women and Development*. Rural Development Committee, Cornell University, Bibliography Series No. 2.

Freyfogle, E. T. 2001. *The New Agrarianism: Land, Culture, and the Community of Life*, Washington DC, Island Press.

Friedman, M and Calixte, S. L. (eds) 2009. *Mothering and blogging: The radical act of the mommyblog*. Bradford, ON, Demeter Press.

Guthman, J. 2008. Bringing good food to others: Investigating the subjects of alternative food practice. *Cultural Geographies*, 15(4), 431-447.

Guthman, J. 2012. *Weighing In: Obesity, Food Justice, and the Limits of Capitalism*. Berkeley, University of California Press.

Hayes, S. 2010. *Radical Homemakers: Reclaiming Domesticity from a Consumer Culture*. Richmondville, NY, Left to Write Press.

Lister, N-M. 2007. Placing food: Toronto's edible landscape.  In J. Knechtel (ed.), *FOOD: Alphabet city series*. Cambridge, MA: MIT Press.

Lyson, T. A. 2004. *Civic Agriculture: Reconnecting Farm, Food, and Community*. Lebanon, NH, Tufts University Press / University Press of New England.

Marsh, R., 1998. Building on traditional gardening to improve household food security. In J. L. Albert, (ed.) *Food, Nutrition and Agriculture*. FAO Food and Nutrition Division, Rome. [http://www.fao.org/docrep/X0051T/X0051T00.htm]

Perfecto, I. 1992. Pesticide exposure of farm workers and the international connection. In *Race and the Incidence of Environmental Hazards*, Edited by B. Bryant and P. Mohai. Westview Press, pp. 177-203.

Pudup, M. B. 2008. It takes a garden: Cultivating citizen-subjects in organized garden projects. *Geoforum*, 39(3), 1228-1240.

Ristau, J. 2011. How to grow a healthy local food system. *On the Commons*. Available at: http://onthecommons.org/how-grow-healthy-local-food-system [accessed 20 April 2012].

Schurman, R. and Munro, W. A. 2010. *Fighting for the Future of Food: Activists Versus Agribusiness in the Struggle over Biotechnology*. University of Minnesota Press, Minneapolis, MN.

TCGN. 2012. *Toronto Community Garden Network*. Available at: http://www.
tcgn.ca/wiki/wiki.php?n=TCGN.FrontPage [accessed: 12 April 2012].

Warner, S. B. 1987. *To Dwell Is to Garden: A History of Boston's Community
Gardens*. Boston, Northeastern University Press.

Wavell, A. P. W. 1945. *Other Men's Flowers: An Anthology of Poetry*. New York,
Putnam.

# Thematic Tabs for Chapter 4

*Body*

*Colonial*

*Race*

*Science*

*Women*

# Chapter 4

# Ancient Dietary Wisdom for Tomorrow's Children

## Sally Fallon Morell

**Editors' Note:** In many ways this chapter might be read more as a narrative to be analyzed than an analytical contribution to 'doing nutrition differently.' We say this for a number of different reasons – the chapter was not written specifically for the volume, but actually many years before its conception, and perhaps because of this Fallon's writing connects up with some of the thematic tabs of the book – e.g., *race* and *colonial* – in ways that have been considered problematic by others in the volume (see Harris, this volume). Still, Fallon's work has had significant influence in challenging the hegemonic approach to dietary advice in the US in recent years – e.g., for 'radical homemakers' connecting up with our *women* tab (see Cadieux this volume). By including her chapter here we both acknowledge this influence and move to examine more closely some of the underlying themes evident in her writing (*colonial*, *race*, *science*, and *women*). We do this in order to explore how the successes of Fallon and the Weston A. Price Foundation might better contribute to a decolonized dietary and nutrition practice.

More than sixty years ago, a Cleveland dentist named Weston A. Price decided to embark on a series of unique investigations that would engage his attention and energies for the next ten years. Possessed of an inquiring mind and a spiritual nature, Price was disturbed by what he found when he looked into the mouths of his patients. Rarely did an examination of an adult client reveal anything but rampant decay, often accompanied by serious problems elsewhere in the body such as arthritis, osteoporosis, diabetes, intestinal complaints and chronic fatigue. (They called it neurasthenia in Price's day.) But it was the dentition of younger patients that gave him most cause for concern. He observed that crowded, crooked teeth were becoming more and more common, along with what Price called "facial deformities" – overbites, narrowed faces, underdevelopment of the nose, lack of well-defined cheekbones and pinched nostrils. Such children invariably suffered from one or more complaints that sound all too familiar to mothers of the 1990s: frequent infections, allergies, anemia, asthma, poor vision, lack of coordination, fatigue and behavioral problems. Price did not believe that such "physical degeneration" was God's plan for mankind. He was rather inclined to believe that the creator intended physical perfection for all human beings, and that children should grow up free of ailments.

Price's bewilderment gave way to a unique idea. He would travel to various isolated parts of the earth where the inhabitants had no contact with "civilization" to study their health and physical development. His investigations took him to isolated Swiss villages and a windswept island off the coast of Scotland. He studied traditional Eskimos, Indian tribes in Canada and the Florida Everglades, South Seas islanders, Aborigines in Australia, Maoris in New Zealand, Peruvian and Amazonian Indians and tribesmen in Africa. These investigations occurred at a time when there still existed remote pockets of humanity untouched by modern inventions; but when one modern invention, the camera, allowed Price to make a permanent record of the people he studied. The photographs Price took, the descriptions of what he found and his startling conclusions are preserved in a book considered a masterpiece by many nutrition researchers who followed in Price's footsteps: *Nutrition and Physical Degeneration.* Yet this compendium of ancestral wisdom is all but unknown to today's medical community and modern parents.

*Nutrition and Physical Degeneration* is the kind of book that changes the way people view the world. No one can look at the handsome photographs of so-called primitive people – faces that are broad, well-formed and noble – without realizing that there is something very wrong with the development of modern children. In every isolated region he visited, Price found tribes or villages where virtually every individual exhibited genuine physical perfection. In such groups, tooth decay was rare and dental crowding and occlusions – the kind of problems that keep American orthodontists in yachts and vacation homes – nonexistent. Price took photograph after photograph of beautiful smiles, and noted that the natives were invariably cheerful and optimistic. Such people were characterized by "splendid physical development" and an almost complete absence of disease, even those living in physical environments that were extremely harsh.

The fact that "primitives" often exhibited a high degree of physical perfection and beautiful straight white teeth was not unknown to other investigators of the era. The accepted explanation was that these people were "racially pure" and that unfortunate changes in facial structure were due to "race mixing." Price found this theory unacceptable. Very often the groups he studied lived close to racially similar groups that had come in contact with traders or missionaries, and had abandoned their traditional diet for foodstuffs available in the newly established stores – sugar, refined grains, canned foods, pasteurized milk and devitalized fats and oils – what Price called the "displacing foods of modern commerce." In these peoples, he found rampant tooth decay, infectious illness and degenerative conditions. Children born to parents who had adopted the so-called "civilized" diet had crowded and crooked teeth, narrowed faces, deformities of bone structure and reduced immunity to disease. Price concluded that race had nothing to do with these changes. He noted that physical degeneration occurred in children of native parents who had adopted the white man's diet; while mixed race children whose parents had consumed traditional foods were born with wide handsome faces and straight teeth.

The diets of the healthy "primitives" Price studied were all very different: In the Swiss village where Price began his investigations, the inhabitants lived on rich dairy products – unpasteurized milk, butter, cream and cheese – dense rye bread, meat occasionally, bone broth soups and the few vegetables they could cultivate during the short summer months. The children never brushed their teeth – in fact their teeth were covered in green slime – but Price found that only about one percent of the teeth had any decay at all. The children went barefoot in frigid streams during weather that forced Dr. Price and his wife to wear heavy wool coats; nevertheless childhood illnesses were virtually nonexistent and there had never been a single case of TB in the village. Hearty Gallic fishermen living off the coast of Scotland consumed no dairy products. Fish formed the mainstay of the diet, along with oats made into porridge and oatcakes. Fish heads stuffed with oats and chopped fish liver was a traditional dish, and one considered very important for children. The Eskimo diet, composed largely of fish, fish roe and marine animals, including seal oil and blubber, allowed Eskimo mothers to produce one sturdy baby after another without suffering any health problems or tooth decay. Well-muscled hunter-gatherers in Canada, the Everglades, the Amazon, Australia and Africa consumed game animals, particularly the parts that "civilized" folk tend to avoid – organ meats, glands, blood, marrow and particularly the adrenal glands – and a variety of grains, tubers, vegetables and fruits that were available. African cattle-keeping tribes like the Masai consumed no plant foods at all – just meat, blood and milk. South Seas islanders and the Maori of New Zealand ate seafood of every sort – fish, shark, octopus, shellfish, sea worms – along with pork meat and fat, and a variety of plant foods including coconut, manioc and fruit. Whenever these isolated peoples could obtain seafoods they did so – even tribes living high in the Andes. These groups put a high value on fish roe, which was available in dried form in the most remote Andean villages. Insects were another common food, in all regions except the Arctic. The foods that allow people of every race and every climate to be healthy are whole natural foods – meat with its fat, organ meats, whole milk products, fish, insects, whole grains, tubers, vegetables and fruit – not newfangled concoctions made with white sugar, refined flour and rancid and chemically altered vegetable oils.

Price took samples of native foods home with him to Cleveland and studied them in his laboratory. He found that these diets contained at least four times the minerals and water soluble vitamins – vitamin C and B complex – as the American diet of his day. Price would undoubtedly find a greater discrepancy in the 1990s due to continual depletion of our soils through industrial farming practices. What's more, among traditional populations, grains and tubers were prepared in ways that increased vitamin content and made minerals more available – soaking, fermenting, sprouting and sour leavening.

It was when Price analyzed the fat soluble vitamins that he got a real surprise. The diets of healthy native groups contained at least ten times more vitamin A and vitamin D than the American diet of his day! These vitamins are found only in animal fats – butter, lard, egg yolks, fish oils and foods with fat-rich cellular membranes like liver and other organ meats, fish eggs and shell fish.

Price referred to the fat soluble vitamins as "catalysts" or "activators" upon which the assimilation of all the other nutrients depended – protein, minerals and vitamins. In other words, without the dietary factors found in animal fats, all the other nutrients largely go to waste.

Price also discovered another fat soluble vitamin that was a more powerful catalyst for nutrient absorption than vitamins A and D. He called it "Activator X" (now believed to be vitamin K2). All the healthy groups Price studied had the X Factor in their diets. It could be found in certain special foods which these people considered sacred – cod liver oil, fish eggs, organ meats and the deep yellow Spring and Fall butter from cows eating rapidly growing green grass. When the snows melted and the cows could go up to the rich pastures above their village, the Swiss placed a bowl of such butter on the church altar and lit a wick in it. The Masai set fire to yellow fields so that new grass could grow for their cows. Hunter-gatherers always ate the organ meats of the game they killed – often raw. Liver was held to be sacred by many African tribes. The Eskimos and many Indian tribes put a very high value on fish eggs.

The therapeutic value of foods rich in the X Factor was recognized during the years before the Second World War. Price found that the action of "high vitamin" Spring and Fall butter was nothing short of magical, especially when small doses of cod liver oil were also part of the diet. He used the combination of high vitamin butter and cod liver oil with great success to treat osteoporosis, tooth decay, arthritis, rickets and failure to thrive in children.

Other researchers used such foods very successfully for the treatment of respiratory diseases such as TB, asthma, allergies and emphysema. One of these was Francis Pottenger whose sanatorium in Monrovia, California served liberal amounts of liver, butter, cream and eggs to convalescing patients. He also gave supplements of adrenal cortex to treat exhaustion.

Dr. Price consistently found that healthy "primitives", whose diets contained adequate nutrients from animal protein and fat, had a cheerful, positive attitude to life. He noted that most prison and asylum inmates have facial deformities indicative of pre-natal nutritional deficiencies.

Like Price, Pottenger was also a researcher. He decided to perform adrenalectomy on cats and then fed them the adrenal cortex extract he prepared for his patients in order to test its effectiveness. Unfortunately most of the cats died during the operation. He conceived of an experiment in which one group of cats received only raw milk and raw meat, while other groups received part of the diet as pasteurized milk or cooked meat. He found that only those cats whose diet was totally raw survived the adrenalectomy and as his research progressed, he noticed that only the all-raw group continued in good health generation after generation – they had excellent bone structure, freedom from parasites and vermin, easy pregnancies and gentle dispositions. All of the groups whose diet was partially cooked developed "facial deformities" of the exact same kind that Price observed in human groups on the "displacing foods of modern commerce" – narrowed faces, crowded jaws, frail bones and weakened ligaments. They were plagued with parasites, developed all manner of diseases and had difficult pregnancies. Female cats became aggressive while the males became docile. After just three generations, young animals died before reaching adulthood and reproduction ceased.

The results of Pottenger's cat experiments are often misinterpreted. They do not mean that humans should eat only raw foods – humans are not cats. Part of the diet was cooked in all the healthy groups Price studied. (Milk products, however, were almost always consumed raw.) Pottenger's findings must be seen in the context of the Price research and can be interpreted as follows: When the human diet produces "facial deformities" – the progressive narrowing of the face and crowding of the teeth – extinction will occur if that diet is followed for several generations. The implications for western civilization – obsessed as it is with refined, highly sweetened convenience foods and low-fat items – is profound.

The research of Weston Price is not so much misinterpreted as ignored. In a country where the entire orthodox health establishment condemns saturated fat and cholesterol from animal sources, and where vending machines have become a fixture in our schools, who wants to hear about a peripatetic dentist who warned about the dangers of sugar and white flour, who thought kids should take cod liver oil and who believed that butter was the number one health food?

The irony is that as Price becomes more and more forgotten, more and more research appears in the scientific literature proving he was right. We now know that vitamin A is essential for the prevention of birth defects, for growth and development, for the health of the immune system and the proper functioning of all the glands. Scientists have discovered that the precursors to vitamin A – the carotenes found in plant foods – cannot be converted to true vitamin A by infants and children. They must get their vital supply of this nutrient from animal fats. Yet orthodox nutritional pundits are now pushing low-fat diets for children. Neither can diabetics and people with thyroid conditions convert carotenes to the fat soluble form of vitamin A – yet diabetics and people with low energy are told to avoid animal fats.

The scientific literature tells us that vitamin D is needed not only for healthy bones, and optimal growth and development, but also to prevent colon cancer, MS and reproductive problems.

Cod liver oil is an excellent source of vitamin D. Cod liver oil also contains special fats called EPA and DHA. The body uses EPA to make substances that help prevent blood clots, and that regulate a myriad of biochemical processes. Recent research shows that DHA is essential to the development of the brain and nervous system. Adequate DHA in the mother's diet is necessary for the proper development of the retina in the infant she carries. DHA in mother's milk helps prevent learning disabilities. Cod liver oil and foods like liver and egg yolk supply this essential nutrient to the developing fetus, to nursing infants and to growing children.

Butter contains both vitamin A and D, as well as other beneficial substances. Conjugated linoleic acid in butterfat is a powerful protection against cancer. Certain fats called glycospingolipids aid digestion. Butter is rich in trace minerals, and naturally yellow Spring and Fall butter contains the X factor.

Saturated fats from animal sources – portrayed as the enemy – form an important part of the cell membrane; they protect the immune system and enhance the utilization of essential fatty acids. They are needed for the proper development of the brain and nervous system. Certain types of saturated fats provide quick energy and protect against pathogenic microorganisms in the intestinal tract; other types provide energy to the heart.

Cholesterol is essential to the development of the brain and nervous system of the infant, so much so that mother's milk is not only extremely rich in the substance, but also contains special enzymes that aid in the absorption of cholesterol from the intestinal tract. Cholesterol is the body's repair substance; when the arteries are damaged because of weakness or irritation, cholesterol steps in to patch things up and prevent aneurysms. Cholesterol is a powerful antioxidant, protecting the body from cancer; it is the precursor to the bile salts, needed for fat digestion; from it the adrenal hormones are formed, those that help us deal with stress and those that regulate sexual function.

The scientific literature is equally clear about the dangers of polyunsaturated vegetable oils – the kind that are supposed to be good for us. Because polyunsaturates are highly subject to rancidity, they increase the body's need for vitamin E and other antioxidants. (Canola oil, in particular, can create severe vitamin E deficiency.) Excess consumption of vegetable oils is especially damaging to the reproductive organs and the lungs – both of which are sites for huge increases in cancer in the US. In test animals, diets high in polyunsaturates from vegetable oils inhibit the ability to learn, especially under conditions of stress; they are toxic to the liver; they compromise the integrity of the immune system; they depress the mental and physical growth of infants; they increase levels of uric acid in the blood; they cause abnormal fatty acid profiles in the adipose tissues; they have been linked to mental decline and chromosomal damage; they accelerate aging. Excess consumption of polyunsaturates is associated with increasing rates of cancer, heart disease and weight gain; excess use of commercial vegetable oils interferes with the production of prostaglandins – localized tissue hormones – leading to an array of complaints such as autoimmune diseases, sterility and PMS. Vegetable oils are more toxic when heated. One study reported that polyunsaturates turn to varnish in the intestines. A study by a plastic surgeon found that women who consumed mostly vegetable oils had far more wrinkles than those who consumed traditional animal fats.

When polyunsaturated oils are hardened to make margarine and shortening by a process called hydrogenation, they deliver a double whammy of increased cancer, reproductive problems, learning disabilities and growth problems in children.

The vital research of Weston Price remains largely forgotten because the importance of his findings, if recognized by the general populace, would bring down America's largest industry – food processing and its three supporting pillars – refined sweeteners, white flour and vegetable oils. Representatives of this industry have worked behind the scenes to erect the huge edifice of the "lipid hypothesis" – the untenable theory that saturated fats and cholesterol cause heart disease and cancer. All one has to do is look at the statistics to know that it isn't true. Butter consumption at the turn of the century was eighteen pounds per person per year, and the use of vegetable oils almost nonexistent, yet cancer and heart disease were rare. Today butter consumption hovers just above four pounds per person per year while vegetable oil consumption has soared – and cancer and heart disease are endemic.

What the research really shows is that both refined carbohydrates and vegetable oils cause imbalances in the blood and at the cellular level that lead to an increased tendency to form blood clots, leading to myocardial infarction. This kind of heart disease was virtually unknown in America in 1900. Today it has reached epidemic levels. Atherosclerosis, or the buildup of hardened plague in the artery walls, cannot be blamed on saturated fats or cholesterol. Very little of the material in this plaque is cholesterol, and a 1994 study appearing in the Lancet showed that almost three quarters of the fat in artery clogs is unsaturated. The "artery clogging" fats are not animal fats but vegetable oils.

Built into the whole cloth of the lipid hypothesis is the postulate that the traditional foods of our ancestors – the butter, cream, eggs, liver, meat and fish eggs that Price recognized were necessary to produce "splendid physical development" – are bad for us. A number of stratagems have served to imbed this notion in the consciousness of the people, not the least of which was the National Cholesterol Education Program (NCEP), during which your tax dollars paid for a packet of "information" on cholesterol and heart disease to be sent to every physician in America. As the American Pharmaceutical Association served on the coordinating committee of this massive program, it is not surprising that the packet instructed physicians on ways to test serum cholesterol levels, and what drugs to prescribe for patients whose cholesterol levels put them in the "at risk" category – defined arbitrarily as anyone over 200 mg/dl, the vast majority of the adult population. Physicians received instruction on the "prudent diet," low in saturated fat and cholesterol, for "at risk" Americans, even though studies indicated that such diets did not offer any significant protection against heart disease. They did, however, increase the risk of death from cancer, intestinal diseases, accidents, suicide and stroke. A specific recommendation contained in the NCEP information packet was the replacement of butter with margarine.

In 1990, two generations after Weston Price conceived of studying isolated non-industrialized people as a way of learning how to confer good health on our children, the National Cholesterol Education Program recommended the "prudent diet" for all Americans above the age of two. The advantage of such a diet is supposed to be reduced risk of heart disease in later life – even though not a single study has shown such an hypothesis to be tenable. What the scientific literature does tell us is that low fat diets for children, or diets in which vegetable oils have been substituted for animal fats, result in failure to thrive – failure to grow tall and strong – as well as learning disabilities, susceptibility to infection and behavioral problems. Teenage girls who adhere to such a diet risk reproductive problems. If they do manage to conceive, their chances of giving birth to a low birth weight baby, or a baby with birth defects, are high.

Compared to this folly, the wisdom of the so-called primitive in regards to ensuring the health of his children has inspired the awe of Weston Price and all who have read his book. Again and again he found that tribal groups – especially those in Africa and the South Pacific – fed special foods to young men and women before conception, to women during pregnancy and lactation, and to children during their growing years. When he tested these foods – things like liver, shellfish, organ meats and bright yellow butter – he found them to be extremely rich in the "fat-soluble activators" – vitamins A, D and the X Factor. Special soaked grain preparations of high mineral content – particularly millet and quinoa – were fed to lactating women to increase milk supply.

Price also discovered that many tribes practiced the spacing of children in order to allow the mother to recover her nutrient stores and to ensure that subsequent children would be as healthy as the first. They did this by a system of multiple wives, or in the case of monogamous cultures, deliberate abstinence. Three years was considered the minimum time necessary between children to the same mother – anything less brought shame on the parents and the opprobrium of the village.

The education of the young in these tribal groups included instruction in dietary wisdom as a way of ensuring the health of future generations and the continuance of the tribe in the face of the constant challenge of finding food, and defending the group against warring neighbors.

Modern parents, living in times of peace and abundance, face an altogether different challenge, one of discrimination and cunning. For they must learn to discriminate between hyperbole and truth when it comes to choosing foods for themselves and their family; and to practice cunning in protecting their children from those displacing products of modern commerce that prevent the optimal expression of their genetic heritage – foodstuffs made of sugar, white flour, vegetable oils and products that imitate the nourishing foods of our ancestors – margarine, shortening, egg replacements, meat extenders, fake broths, ersatz cream, processed cheese, factory farmed meats, industrially farmed plant foods, protein powders, and packets of stuff that never spoils.

For a future of healthy children – for any future at all – we must turn our backs on the dietary advice of sophisticated medical orthodoxy and return to the food wisdom of our so-called primitive ancestors, choosing traditional whole foods that are organically grown, humanely raised, minimally processed and above all not shorn of their vital lipid component.

When offspring are properly spaced, and care given to the diet of both parents before conception, and to the children during their period of growth and development, all children in the family can be blessed with the kind of good health that allows them a carefree childhood; and the energy and intelligence they need to put their adult years to best and highest use.

*Editor's Note: Please see Harris, this volume, for an analytical look at the work of Sally Fallon Morell and the Weston A. Price Foundation.*

# Thematic Tabs for Chapter 5

*Colonial*

*Nature and Structure*

Chapter 5

# Nutritional and Cultural Transitions in Alaska Native Food Systems: Legacies of Colonialism, Contested Innovation, and Rural-Urban Linkages

David V. Fazzino II and Philip A Loring

**Editors' Note:** This chapter centers on the thematic tabs of *colonial*, *nature*, and *structure*. One of Fazzino and Loring's overall contributions is to encourage more sensitivity and more specificity (and less uniformity) in how nutrition is defined and practiced together with indigenous communities. In the chapter, Fazzino and Loring examine the relationship of Alaskan and federal politics and policies in restricting access to Alaska Natives' traditional foods, focusing on the impacts of fisheries management under the Yukon River Salmon Agreement. Fazzino and Loring's work might be understood as work towards *nutrition justice* as their arguments line up with those of the food justice movement, recognizing the need to the need to examine patterns of social and economic inequity, and the policies that create or reinforce this inequity, in order to understand how nutrition could be done differently.

## Introduction

We argue that it is insufficient to envision health as solely the outcome of individual agency within an idealized context of a free market that offers equal access to healthful and preferred sources of food and medicine. Dietary health outcomes are equally linked to the availability and affordability of quality food options in any given community, as well as to the quality of information available to people regarding food and nutrition (Gittelsohn and Sharma 2009). Hence, as with environmental health in general (Beckfield and Krieger 2009; Poundstone et al. 2004), the locus of nutritional interventions should not solely repeat sound bites for personal responsibility, but should also consider structural and societal impediments to achieving healthy lifestyles. Too, it is not sufficient to identify interventions solely through reference to biomedical metrics but in unison with local communities who can in effect "indigenize," following Sahlins (1999), nutrition. Central to this argument is that the category of so-called "traditional" or "country" foods in Alaska are not limited to those foods consumed prior to the arrival of Russian and Euroamerican immigrants to the region, but rather involves a constantly shifting landscape of country foods, with the emphasis being on flexibility and diversity of food options rather than a menu of foods and subsistence practices that occurred prior to some arbitrary event or legislation (Loring and Gerlach 2010b).

As will become evident throughout the course of this essay, nutrition is not something that can or should be uniformly regulated at the State level; rather, we argue that optimum health outcomes are only possible when State policies create opportunities for indigenous initiatives to promote health and wellness to thrive (c.f. Loring and Duffy forthcoming). Here, we utilize the Declaration on the Rights of Indigenous Peoples, examining the relationship of Alaskan and federal politics in curbing access to the traditional or "country foods" of Alaska Native peoples, with specific emphasis on the impacts of fisheries management under the Yukon River Salmon Agreement on indigenous peoples of Alaska.


**Country Foods: Continuity and Change**

For millennia, Alaska Native foodways were based almost entirely on locally harvested "country" foods (Usher 1976), including (depending on region) sea mammals, ungulates such as caribou and moose, freshwater and saltwater fish, seasonally available waterfowl, formal and informal gardens, berries, and other plant resources. Long-standing patterns of land-use and landscape features demarcated general but flexible boundaries around each tribal group's foodshed (Loring 2007). These traditional foodways continue to connect Alaska Natives in physical and cultural ways to the land and wildlife through activities such as food sharing and food preparation, and the use of traditional travel routes, harvest sites, and camps of modern and historical significance. Today, however, country foods make up only a fraction of the entire diet for most Alaska Natives, more than half of whom live in urban centers such as Anchorage (Martin et al. 2008). While country foods remain preferred by most Alaska Natives, a variety of socioeconomic and environmental circumstances limit the ability of many to subsist primarily or even partially on country foods (Loring and Gerlach 2009).

Still, for rural residents, at least 60% of the contribution of wild foods to Alaska Native diets throughout the state comes from fish species, and while outmigration from rural to urban areas is continuing, so are anecdotal reports of the sharing of country foods between rural and urban areas (Lee 2003). In addition to their basic caloric and nutritive contributions, it is important to note that salmon and other game species play iconic, "cultural keystone" roles in many of these communities (Garibaldi and Turner 2004), including for instance moose and caribou in the Interior, and whale and walrus on the coast. Hence, different country foods are understood to play various roles in the health of these people and their communities (Loring and Gerlach 2009).

Challenges for the rural hunter or fisher are very different than they were even two decades ago: the current economy of many rural residents and particularly

Alaska Natives is often described as 'mixed-subsistence' wherein money is earned in order to provide for the necessary supplies and tools for hunting and fishing. These tools include gas-driven vehicles such as boats, all-terrain vehicles (ATVs) and snow-machines, as well as the requisite fuel and parts for their maintenance. The rising costs of fuel alone can be prohibitive; time is also in a necessary-but-scarce supply as many are forced to make trade-off decisions between spending time on the land or on earning wages necessary to underwrite the cost of the hunt, keep fuel in the stoves, and keep cupboards full as these days few households are entirely dependent on country foods.

This concept of 'mixed-subsistence' economy, implying that there are some activities which qualify as subsistence but others that do not, illustrates the general failure of 'subsistence' as currently defined and operationalized in state and federal law and regulation to fully capture the complexity and character of rural life (Gerlach et al. forthcoming). Despite the narrowly and rigidly construed definition of subsistence as customary and traditional practices, found in the Alaska Native Claims Settlement Act (ANCSA), rural peoples rarely divide their lives into activities that are for-subsistence or otherwise. The exception, however, is in environmental and political discourses where invocation of the term is understood as useful for mobilizing political will or enacting specific statutory protections in the attempt to control both the meaning and use of Alaska's landscapes (Behnke 2002: 152).

## The Nutrition Transition

Foodways across the state are transitioning in multiple ways that are still poorly documented or understood. There has been a marked movement away from "country foods" in rural communities as more food is purchased from the store (Bersamin et al. 2007, Kuhnlein et al. 2004). In general, food production and procurement options are quite limited in the villages, by a lack of employment opportunities, by the costs and challenges of transport to and from urban supply centers, and by lack of agricultural and manufacturing infrastructure (Colt et al. 2003; Martin et al. 2008). Driven by necessity as well as by expedience, this "nutrition transition" is widespread across the North American Arctic (Kuhnlein et al. 2004; Huntington 1992), though the extent varies significantly from community to community. Some rural communities in Alaska enjoy a wide variety of readily available fish and game, maintaining a strong preference for a utilization of traditional foods, while others cope with near-food deserts. Climate change, competition from tourist and sport hunters/fishers, lands and natural resources development, and a rigid wildlife management regime at both state and federal levels can all significantly confound the viability of wild foods as a source of economic security and stability (Huntington 1992).

This transition has not been without its caveats, both obvious and hidden, and a variety of ongoing research is working to quantify the economic, nutritional, even psychological and psychosocial outcomes that continue to emerge (Bersamin et al. 2007; Graves 2004). We can say, without claims of clear or straightforward causality, that contemporary Alaska Native populations suffer an unprecedented burden of physical, mental, and social health syndromes and diseases. These include near-epidemic increases in cancer, coronary heart disease, type 2 diabetes, depression and alcoholism. These are all trends with established environmental and dietary health components, and are trends that were anticipated by the earlier works of Price (1939) and Hurwitz (1977).

This nutrition transition is also not without antecedents as earlier attempts to alter the food system attest. The perceptions of poverty, food shortage, and impending famine, especially for the state's rural, or 'Native' residents, have been used repeatedly in Alaska as means to purport (or enact) a particular political ideology. Famine in Northwest Alaska in the late 1800s, both perceived and real, was leveraged to forward Presbyterian Minister Sheldon Jackson's agenda for the import of reindeer herding to the purportedly imperiled Eskimo communities as a mechanism of economic aid, education, and civilization (Gerlach 1996; Hinckley 1966; Willis 2006).

Agricultural outreach and economic development programs by the Bureau of Indian Affairs (BIA), the Alaska Native Service (ANS), the Cooperative Extension Service (CES), and others all similarly shared goals of saving the Native from their "situation" (Loring and Gerlach 2010; Willis 2006). The perceptions of economic need and uncertainty were also highly-advertised benefits touted for both the Indian Reorganization Act (IRA) and the Alaska Native Claims Settlement Act (ANCSA) (Case and Mitchell 2003; Sacks 1995), legislation that has left an indelible mark on the political landscape of the state, with mixed outcomes at best (Berger 1995).

### Contemporary Challenges: Yukon River Salmon

Food crises have been publicized recently for Alaskans (Fazzino and Loring 2009), who face a number of challenges in regards to their food systems that will need to be addressed in order to ensure the long-term viability and sustainability of rural communities. Regional vulnerabilities to external market shifts in the price or availability of imported foods and fuel are just two poignant examples (AKDHSS 2008). In addition, contemporary challenges, such as recent salmon fishery failures on the Yukon River, display how the nutritional ecology of rural Alaska links with the cumulative impacts of climate change, and conflicting uses and management agendas regarding wild fish and game.

In 2009, for example, a fisheries failure of the Chinook (king) salmon fishery on the Yukon River highlighted ongoing food insecurity challenges for Alaskans as well as the need for more comprehensive and proactive policy measures for reducing the vulnerability of Alaska food systems. Alaska Natives living in communities along the river, and especially in upper Yukon River watersheds faced unprecedented salmon fisheries closures and country food shortfalls in 2009, followed by another difficult year fishing in 2010 (Loring and Gerlach 2010). Mandates at the state and federal level, however, situate management of salmon on the Yukon in a secular context of single-species conservation rather than food security or environmental sustainability, and the management actions taken in 2009 were as a result considered a success despite regional food shortages, while the 2010 year, in which many more people caught fish, was considered a management failure (Loring and Gerlach 2010).

Yukon River salmon fisheries fall within the international jurisdiction of the Pacific Salmon Treaty (PST) signed by the US and Canada in 1985. The PST was drafted to address questions of fairness in the interception of salmon originating in one nation by the fisheries of the other nation. The PST covers salmon stocks in Alaska as well as in Idaho, Washington, Oregon, Yukon Territory and British Columbia. The primary mandate of the PST is conservation, via both the elimination of overfishing as well as the restoration of degraded salmon populations. Most of the stipulations of the PST deal with ocean fisheries intercepting salmon hundreds of miles from their river of origin; trans-boundary river scenarios like the Yukon are noted, but the PST does not specifically address Yukon River stocks. This was addressed in Yukon River Salmon Agreement (YRA), which was established in 2001 following a devastatingly poor king salmon run in 2000. The YRA is an international commitment to the conservation, restoration, and sustainable harvest of Yukon salmon. Both the United States and Canada agreed to manage their salmon fisheries to ensure enough spawning salmon reach their spawning grounds. With populations of Yukon River salmon at distressing lows when the YRA was established, the agreement primarily addressed the need to rebuild Canadian-origin stocks back to 'historical' levels. Ideally, however, objectives are set to not just maintain a viable salmon population but to maximize that population for a surplus of harvestable salmon by subsistence and commercial fishers in the US and Canada. Conservation remains the first priority, however, and both countries have agreed to decrease their Total Allowable Catch (TAC), or even close fisheries outright in order to protect spawning escapement in years of low runs.

As the PST and YRA are primarily concerned with salmon conservation, they are written in the language of the biology of the fishery. Down-river and up-river, for instance, reflect genetic sub-populations of salmon rather than civic or geographic units. These concepts thus codify in law the state of art for scientific knowledge about the salmon population itself; at the time the YRA was established, 50 percent of the salmon population was believed to spawn after entering Canada via the main-stem of the river; these are called "up-river" salmon. With land use, environmental, and climatic change, however, this number is, or at least should be suspect (see Loring and Gerlach 2010), but it would take a change in the international treaty to adjust effectively to changes in the environment and salmon behavior. One result has been that the interests of communities of the upper Yukon River basin in Alaska, commonly called the Yukon Flats villages, are obscured by the interests of salmon escapement along the main stem of the Yukon to Canada, with Yukon Flats communities more vulnerable to salmon shortfalls and management errors (Loring and Gerlach 2010).

While the PST and YRA ensure protections for the salmon themselves and ensure harvest-sharing between US and Canada in "normal" years, there are few protections for Alaskans to salmon shortfalls on the river. In the US, participants in the commercial portion of the salmon fisheries do, however, obtain some protection from the Magnuson Stevens Fisheries Conservation and Restoration Act (MSFA), 16 U.S.C. § 1861 section 213A, which contains language that empowers the US Department of Commerce to declare a "fishery failure" and recommend that recovery grant funding be made available to the petitioning organization (i.e., the state). Yet, under MSFA, subsistence and "personal-use" fishers, who by Alaska's state constitution have priority over commercial users, have no protections in the case of a failure or disaster. In response to the 2009 king salmon fisheries closures, Alaska governor Sean Parnell successfully petitioned the US government under the MSFA. Ultimately, a sum of five million dollars was made available for relief, though this relief was limited in distribution to those who could document loss in the commercial fishery, both with commercial fishing permits from 2009 and documentation of catch from a prior year (Pacific States Marine Fisheries Commission 2010). Impacts on subsistence fisheries, including the economic costs associated with not fishing (purchasing more food at the store) and enforcement (the cost of pulling fish-wheels from the river), and the more difficult to track sociocultural and health impacts, are far from being fully understood and have no mechanism for drawing similar compensation.

## Confronting Crisis and Applying The Declaration on the Rights of Indigenous Peoples

We argue that the current management regime of the Yukon River salmon fundamentally contradicts aspects of The Declaration on the Rights of Indigenous Peoples. The Declaration on the Rights of Indigenous Peoples establishes a normative framework for the manner that States must interact with indigenous peoples located within the current State's geopolitical boundaries, however the continuing impacts of the YRA should raise pertinent questions concerning the legal and ethical ramifications of both the US and Canada's roles in limiting access to country foods. The passage and adoption of the Declaration by the majority of States, including Canada indicates that the rights of indigenous peoples are becoming increasingly recognized. Among the Articles of the Declaration on the Rights of Indigenous Peoples which are relevant to continued management practices under the YRA are Articles 3, 4, 11, 23, 24, 26, 29 and 32.

Fundamental to any discussion of the rights of indigenous peoples is the right to self-determination. This right is addressed both directly and indirectly by the Declaration on the Rights of Indigenous Peoples. Article 3 concerns the right to self-determination. Article 4 addresses the right to, "ways and means for financing their autonomous functions." For indigenous peoples relying upon country foods in a mixed subsistence economy, control over and access to fisheries and other subsistence resources is essential to retaining self-determination in a manner consistent with the Declaration on the Rights of Indigenous Peoples.

Article 11 relates to subsistence, noting under Article 11(1), "Indigenous peoples have the right to practice and revitalize their cultural traditions and customs." Further, Article 11(1) does not essentialize indigenous identities as only reflective of past traditions, but rather, "... includes the right to maintain, protect and develop the past, present and future manifestations of their cultures ..." This allows for indigenous peoples, like humans everywhere, the opportunity to innovate, to adopt new technologies and fashion them towards their own ends without fundamentally undermining their right of self-determination (Sahlins 1999).

Indigenous peoples, in many contexts, consider a return to or perpetuation of culturally-acceptable food consumption as an essential component in confronting health and nutritional consequences of nutrition transitions (Ayerza and Coates 2005; Fazzino 2008; Komlos 2003; LaDuke 2006; Nabhan 1989; Ogoye-Ndegwa and Aagaard-Hansen 2006; Pieroni and Quave 2006: 101-128; Pilcher 2005: 235-250; Pottier 1999; Turner 1995). As the move away from traditional foods has had dramatic health impacts on indigenous peoples Article 23 and Article 24 of the Declaration on the Rights of Indigenous Peoples is also applicable in considering traditional food security. In addition access to traditional means of subsistence including fish is an important aspect of both inter-generational and intra-generational knowledge exchange, allowing for both cultural continuity as well as adaptive capacity. These exchanges solidify sense of place and reinforce cultural identity in the face of both internal and external pressures, hence serving as important markers of self and identity in terms of mental health. Article 23 notes that, "Indigenous peoples have the right to determine and develop priorities and strategies for exercising their right to development. In particular, indigenous peoples have the right to be actively involved in developing and determining health, housing and other economic and social programmes affecting them and, as far as possible, to administer such programmes through their own institutions." The exercise of such rights is not considered as contrary to acceptable public health practice (Hassel 2006, Loring and Duffy forthcoming). Further Article 24 states that indigenous peoples have the right to maintain their health practices and "an equal right to the enjoyment of the highest attainable standard of physical and mental health."

In regards to conservation, Article 29(1) of the UNDRIP notes that, "States shall establish and implement assistance programmes for indigenous peoples for such conservation and protection, without discrimination." Here the intent is to balance conservation and management goals or scientific management of resources with the needs of indigenous peoples in mind. Certainly in light of current fisheries management in the Yukon River this is, in Canada, acknowledged, while the impacts of the current management regime in the United States clearly illustrates that this is not the case.

Historical development projects have been foisted upon Alaska Natives. These include a variety of initiatives in agriculture (Loring and Gerlach 2010). Some of these have been adopted and incorporated into contemporary conceptualizations of subsistence amongst Alaska Natives (Loring and Gerlach 2010). Legal definitions of subsistence in Alaska are seemingly static as they refer to the utilization of fish and game and neglect the historical importance of outpost agriculture which has been incorporated into subsistence strategies (Loring and Gerlach 2010). In regards to which activities count as subsistence Article 32(1) of the UNDRIP notes that, "Indigenous peoples have the right to determine and develop priorities and strategies for the development or use of their lands or territories and other resources."

Scientific based management regimes that promise to ensure the sustainability of future resources for the common good of all citizens, nonetheless bypass the continuing needs of indigenous communities, and thus are arguably are in direct contradiction to the intent and spirit of the UN Declaration on the Rights of Indigenous Peoples. One way this could change would be to change how 'sustainability' is being conceived of and who it is designed for, from the single-species mind-set that focuses on 'resources,' to an approach based on food sovereignty and foodshed security (Kloppenberg et al. 1996, Rosset 2008). Inherent to these approaches are ways to not merely reconcile perceived trade-offs between conservation and protection of resources with the rights of Alaska Natives, but solutions that are mutually reinforcing of both goals, by allowing for flexibility and adaptability in their food systems. Foodways diversity has been shown repeatedly to maximize nutrition; rather than lock people in to one or two "traditional" foods that are high in nutrition, with alternatives in a time of shortage or stress being "inferior," the goal should be a dynamic continuum of foods that provide complete nutrition in a biophysical and cultural sense, even during times of stress.

**Concluding Remarks**

The conceptualization of food security at the State level is intimately connected with State goals of stability and acquiescence to powerful entities and less so with supporting multiple cultural interests (Fazzino 2004; Fazzino 2010). Along these lines many definitions of food security provided by federal agencies fail to adequately consider the importance of culturally acceptable foods (Fazzino 2010). This contradicts not only the internationally recognized right to food but also the UN Declaration on the Rights of Indigenous Peoples.

Doing nutrition differently in the Alaska context means moving beyond the discourse of food system vulnerability in the sense of mitigative responses to isolated episodes of "non-normal" conditions, such as food crisis. Rather giving attention to increasing the resilience and flexibility of communities; in other words, efforts to rebuild or enhance food security should involve measures that increase the amount of stress or disturbance that can be comfortably absorbed or accommodated within a food system without the experience of significant strain (Ericksen et al. 2010). We think that the redevelopment of self-reliance into local food system designs is one avenue. Historically, Alaska Native foodways have been resilient to intermitted disruptions and to natural variation in the abundance of fish and game, through diversity and flexibility in the food system (Binford 1978, Loring and Gerlach 2010b). New policy options to enhance local food security that align with the UN Declaration on the Rights of Indigenous Peoples could focus on supporting Alaska Natives in their attempts to rebuild flexibility and diversity into their food system, whether through innovative approaches to fish and game co-management that transcend the single-species approach, or with support for what appears to be an emerging model of highly mobile rural residency, as people move back and forth between "bush" villages and urban centers, whether for jobs, to care for loved ones, or to hunt, fish, celebrate, and teach their children. Too, we must take seriously the call from the Declaration on the Rights of Indigenous Peoples to not essentialize indigenous identities as only reflective of past traditions. Indeed, foods which are considered to be culturally-appropriate foods include not only traditional pre-Contact foods that demonstrate continuity with the past but also those foods which have been indigenized or fashioned by indigenous peoples themselves to perform nourishing functions within their own society. Indeed, as the case of the Yukon River salmon demonstrates indigenous peoples are continually drawn into political, economic and scientific debates concerning natural resources which can fundamentally buttress or erode their well being. Increased recognition by States to fundamental human rights of indigenous peoples under the UNDRIP offers one potential avenue to consider nutrition in a new way.

## References

Alaska Department of Health and Social Services. 2008 Health Risks in Alaska among Adults.

Alaska Department of Health and Social Services. Alaska Behavioral Risk Factor Survey 2007 Annual Report.

Ayerza Jr., Ricardo and William Coates. 2005. *Chia*. Tucson: The University of Arizona Press.

Beckfield, J., and N. Krieger. 2009. Epi + demos + cracy: Linking Political Systems and Priorities to the Magnitude of Health Inequities--Evidence, Gaps, and a Research Agenda. Epidemiologic Reviews 31 (1) (May): 152-177.

Behnke, S. 2002. Alaska's Contested Landscapes and the Subsistence Claims of Alaska Natives. In *Geographical Identities of Ethnic America, 149-173*. Reno: University of Nevada Press.

Berger, Thomas R. 1985. *Village Journey: The Report of the Alaska Native Review Commission*. Hill & Wang: New York.

Bersamin, A., B.R. Luick, E. Ruppert, J.S. Stern, and S. Sidenberg-Cherr. 2007. Diet Quality among Yup'ik Eskimos Living in Rural Communities is Low: The Center for Alaska Native Health Research Pilot Study. *Journal of the American Dietetic Association* 106 (7): 1055-1063.

Binford, Lewis R. 1978. *Nunamuit Ethnoarchaeology.* New York: Academic Press.

Case, David and David C. Mitchell. 2003. *Sold American.* University of Alaska Press: Fairbanks, Alaska.

Colt, Steve, Scott Goldsmith and Anita Wiita. 2003. *Sustainable Utilities in Rural Alaska: Effective Management, Maintenance and Operation of Electric, Water, Sewer, Bulk Fuel, Solid Waste.* Anchorage, AK: Institute of Social and Economic Research, University of Alaska Anchorage.

Ericksen, Polly J., Hans Georg Bohle, and Beth Stewart. 2010. Vulnerability and Resilience in Food Systems. In *Food Security and Global Environmental Change*, 67-77. Earthscan: London, UK.

Fazzino, David. 2010. Whose Food Security?: Confronting Expanding Commodity Production and the Obesity and Diabetes Epidemics. *Drake Law Review* 15(3).

—— 2008. Continuity and Change in Tohono O'odham Food Systems: Implications for Dietary Interventions, *Culture & Agriculture* 30: 38-46.

——2004. The Meaning and Relevance of Food Security in the Context of Current Globalization Trends. *Journal of Land Use & Environmental Law* 19(2): 435-50.

Fazzino, David and Philip A. Loring. 2009. From Crisis to Cumulative Effects: Food Security Challenges in Rural Alaska. NAPA Bulletin 32: 152-177.

Garibaldi, Anne and Nancy Turner. 2004. Cultural Keystone Species: Implications for Ecological Conservation and Restoration, Ecology and Society 9(3) Electronic resource, http://www.ecologyandsociety.org/vol9/iss3/art1/print. pdf (accessed September 13, 2011).

Gerlach, Scott C. 1996. Historical Archaeology and the Early Twentieth Century Reindeer Herding Frontier on the Northern Seward Peninsula, Alaska (pp. 95-98). In *Ublasaun: First Light.* National Park Service.

Gerlach, S. Craig, Philip A. Loring, and Amy M. Turner. Forthcoming. Food Systems, Climate Change, and Community Needs. *In North by 2020*, edited by Hajo Eicken and A.L. Lovecraft. University of Alaska Press, Fairbanks, AK.

Gittelsohn, Joel, and Sangita Sharma. 2009. Physical, Consumer, and Social Aspects of Measuring the Food Environment Among Diverse Low-Income Populations. *American Journal of Preventive Medicine* 36(4) (April): S161-S165.

Graves, K. 2004. Resilience and Adaptation Among Alaska Native Men: Abstract. *International Journal of Circumpolar Health* 63(1): 2004.

Hassel, C.A. 2006. Woodlands Wisdom: a nutrition program interfacing indigenous and biomedical epistemologies. *Journal of Nutrition Education and Behavior* 38(2): 114-120.

Hinckley, Ted C. 1966. The Presbyterian Leadership in Pioneer Alaska. *Journal of American History*. 52(4), 742-756.

Huntington, Henry. 1992. *Wildlife Management and Subsistence Hunting in Alaska*. Seattle: University of Washington Press.

Hurwitz, B. 1977. Subsistence Foods: Physician's Perspective on the d2 Land Proposal, *Alaska Medicine* 19: 60-66.

Kloppenburg, J., J. Hendrickson, and G.W. Stevenson. 1996. Coming into the Foodshed. *Agriculture and Human Values* 13(3): 33-42.

Komlos, John. 2003. Access to Food and the Biological Standard of Living: Perspectives on the Nutritional Status of Native Americans. *The American Economic Review* 93(1): 252-55.

Kuhnlein, V., O. Receveur, R. Soueida and G. M. Egeland, 2004 Arctic Indigenous Peoples Experience the Nutrition Transition with Changing Dietary Patterns and Obesity, *Journal of Nutrition* 134: 1447-1453.

LaDuke, Winona. 2006. *Recovering the Sacred: The Power of Naming and Claiming*. Cambridge, MA: South End Press.

Lee, Molly C. 2003. The Cooler Ring: Urban Alaska Native Women and the Subsistence Debate, *Arctic Anthropology* 30: 3-9.

Loring, Philip A., and Lawrence K. Duffy. Forthcoming. Managing environmental risks: the benefits of a place-based approach. *Remote and Rural Health* 11(3).

Loring, Philip A., S. Craig Gerlach, David E. Atkinson and Maribeth Murray. 2011. Ways to Help and Ways to Hinder: Governance for Effective Adaptation to a Uncertain Climate. *Arctic* 64(1): 73-88.

Loring, Philip A., Gerlach, S. Craig. 2010a. Food Security and Conservation of Yukon River Salmon: Are We Asking Too Much of the Yukon River? *Sustainability* 2(9): 2965-2987.

—— 2010b. Outpost Gardening in Interior Alaska: Food System Innovation and the Alaska Native Gardens of the 1930s through the 1970s. *Ethnohistory* 57(2): 183-199.

—— 2009. Food, Culture, and Human Health in Alaska: An Integrative Health Approach to Food Security. *Environmental Science and Policy* 12(4), pp. 466-478.

Martin, Stephanie, Mary Killorin, Steve Colt. 2008. Fuel Costs, Migration, and Community Viability. Anchorage: Institute of Social and Economic Research. Electronic resource, http://www.alaskapower.org/pdf/Fuelcost_viability_final.pdf (accessed September 13, 2011).

Nabhan, Gary Paul. 1989. *Enduring Seeds: Native American Agriculture and Wild Plant Conservation*. San Francisco: North Point Press.

Ogoye-Ndegwa, Charles and Aagaard-Hansen, Jens. 2006. Dietary and Medicinal Use of Traditional Herbs Among the Luo of Western Kenya. In *Eating and Healing: Traditional Food as Medicine*. Andrea Pieroni and Lisa Leimar Price (eds) pp. 323-343. New York: Food Products Press.

Pacific States Marine Fisheries Commission. 2010. Yukon River Commercial Fishery Disaster Federal Disaster Relief Program (p. 1). Portland, OR:

Pacific States Marine Fisheries Commission. Electronic resource, http://www. psmfc.org/files/October%202010/yukonriversalmondisasterinformation.pdf (accessed September 13, 2010.

Pieroni, Andrea and Cassandra L. Quave. 2006. Functional Foods or Food Medicines? On the Consumption of Wild Plants Among Albanians and Southern Italians in Lucania. In *Eating and Healing: Traditional Food as Medicine*. Andrea Pieroni and Lisa Leimar Price (eds) pp. 101-129. New York: Food Products Press.

Pilcher, Jefery M. 2005. Industrial Tortillas and Folkloric Pepsi: The Nutritional Consequences of Hybrid Cuisines in Mexico. In *The Cultural Politics of Food and Eating, a Reader*. James L Watson and Melissa L Caldwell (eds) pp. 235-250. Malden, MA: Blackwell Publishing.

Pottier, Johan. 1999. *Anthropology of Food: The Social Dynamics of Food Security*. Malden, MA: Polity Press, Blackwell Publishers, Inc.

Poundstone, K. E., S. A. Strathdee, and D. D. Celentano. 2004. The Social Epidemiology of Human Immunodeficiency Virus/Acquired Immunodeficiency Syndrome. *Epidemiologic Reviews* 26 (1) (July 1): 22-35.

Price, Weston A. 1939. *Nutrition And Physical Degeneration: On The Problems of the Western Diet and the Obsession with Nutrients* (Vol. 8). McGraw-Hill Book Company.

Rosset, Peter. 2008. Food Sovereignty and the Contemporary Food Crisis. *Development* 51 (December): 460-463.

Sacks, Jeremy D. 1995. Culture, cash or calories: Interpreting Alaska Native Subsistence Rights. *Alaska Law Review* 12: 247-291.

Sahlins, Marshall. 1999. What Is Anthropological Enlightenment? Some Lessons of the Twentieth Century. *Annual Review of Anthropology* 28:i–xxiii.

Saylor, B., K. Graves, and P. Cochran. 2006. A Resilience-Based Approach to Improving Community Health. *Northwest Public Health* 23(1): 4.

Tweeten, Luther. 1999. The Economics of Global Food Security. *Review of Agricultural Economics* 21(2): 473-488.

Willis, Roxanne R. 2006. *Making Alaska American: Environment and Development in a Foreign Land*. Yale University, History.

Wolfe, R.J. 2004. *Local Traditions and Subsistence: A Synopsis of Twenty-Five Years of Research by the State of Alaska*. Alaska Department of Fish and Game, Division of Subsistence.

# Thematic Tabs for Chapter 6

*Access*

*Emotion*

*Science*

*Structure*

# Chapter 6

# Counseling the Whole Person

### Laura Frank

**Editors' Note:** This chapter connects with the themes of *access, emotion, science,* and *structure.* Frank offers a model for nutritional counseling that begins with a more holistic understanding of an individual's socially embedded lived reality—drawing in matters of socioeconomic constraints, emotion, cultural influences, and religious/spiritual life, among others. In developing this holistic model, Frank re-situates nutrition science, not diminishing its importance, but de-centering its authority amid the multiplex foci necessary for the sincere practice of nutrition.

## Introduction

Food—growing, preparing, eating and sharing—can be one of the greatest sources of human pleasure, health, and connection. For people, it is much more than a source of nutrients that sustain our bodies. It is an expression of creativity, culture and religious belief; it invokes and conveys emotions and memories; and it allows us to nourish ourselves while connecting us to our history and the land that sustains us. Unfortunately, in modern society, food and eating have lost much of that meaning for many people. For those people, many of whom are our clients and patients, eating is an inconvenience rather than a pleasure, and food is an enemy that can make them sick or overweight. Choosing the "right" foods means balancing a confusing array of factors: "superfood" or "supersize"? Food that tastes good, or food as medicine? Convenience foods, or "health" foods? Is good food organic and sustainably produced, or genetically modified nutraceuticals? What does it mean to eat well?

Clearly, although an understanding of normal and therapeutic nutrition is a necessary basis for nutrition counseling, it is only the beginning of what the nutrition counselor needs to know to help clients to find ways to eat that are personally enjoyable, appropriate, and feasible within the context of their lives. The successful nutrition counselor has to be part scientist, part therapist/coach, part cultural anthropologist, part chef/gourmand, part social worker, part fitness advisor—a skill set far broader than that encompassed by the typical training of the nutrition professional.

This chapter is written for professionals who counsel individuals regarding their food choices and eating behavior, as well as students and others interested in how individuals conceptualize their health and make choices that affect their relationship with food and their bodies. As such, the focus is primarily on the counseling process rather than on the information to be conveyed. Although it does discuss theories of health behavior and various counseling techniques and approaches, the intent is to provide this in the context of a more holistic model of counseling that takes into account the client's aesthetic, emotional, cultural, socioeconomic, religious/spiritual, and lifestyle influences on food preferences and choices. The ultimate goal of this counseling model is to facilitate a healthy relationship with food that nourishes the client beyond ensuring an appropriate nutrient profile, and enables the client to select food that is enjoyable, appropriate to his/her health status and goals, and also appropriate to the client's lifestyle and preferences. Other issues related to promoting a healthy relationship with food, such as food access, conscious eating/slow food, and sustainable choices are also addressed. The nutrition counselor is encouraged to evaluate the multiple aspects of food and eating that are important to today's client, and to tailor his/her counseling to the wants and needs of each individual.

### What is Nutrition Counseling?

I have been teaching nutrition counseling to nutrition and dietetics students for nearly two decades, yet this basic question that I ask when the course begins is still a puzzler to many students, regardless of whether they are undergraduates, interns, or practicing dietitians. We have to start by discussing what is *not* nutrition counseling: the "diet instruction" model, in which the patient or client is to be taught the proper way to select, prepare, and consume foods according to his/her medical and nutrient needs, usually based on a medical diagnosis. This model makes several assumptions, including:

- The dietitian is the expert, whose role is to decide what information is important to the client, and provide that information.
- The content of instruction should be standardized according to the diagnosis or identified medical problem.
- The client's role is to receive information and apply it to make changes that the health care providers deem medically necessary.
- The client is obligated to either comply with the recommended changes, or be willing to accept whatever negative consequences will follow from "non-compliance."
- The dietitian successfully completes the diet instruction by providing all of the information that the client "needs to know."

So, how does this differ from nutrition counseling? The counseling model operates on a different set of assumptions:

- Although the dietitian is the expert on nutrition, the client should decide what food-related information is important, useful, and relevant—the role of the dietitian is to help the client to figure that out.
- The information content should be tailored to the client's individual life circumstances as well as the diagnosis.
- The client's role is to collaborate with the dietitian to identify manageable changes that will benefit the client's health and promote a healthy relationship with food.
- The client is obligated to make an educated choice as to which changes will be beneficial and acceptable, and carry out a good faith effort to make those changes with the information and tools provided.
- The dietitian successfully completes the counseling interaction by actively listening to the client, helping the client to set mutually agreeable goals, providing information and tools that are useful and relevant to the individual client, strategizing with the client and ensuring that the client feels competent to implement a mutually agreed-upon plan of action.

Nutrition counseling is client-centered rather than dietitian-centered; it is facilitative and collaborative rather than directive; it is also non-judgmental and respectful of the client's motivations and priorities. Further, the focus of nutrition counseling is less on imparting information, and more on changing behavior.

## Useful Models for Behavior Change

A bit of psychology humor that I heard while I was in graduate school studying counseling has always stuck with me: Q: "How many psychologists does it take to change a light bulb?" A: "Only one, but the bulb has to *want* to change". Silly, yes, but it makes a good point: people change when *they* want to, not when the dietitian wants them to. So, the dietitian's job is to help the client to decide that change is worthwhile and possible, and to provide ideas and information to support the change that the client decides upon.

Changing an individual's nutritional status requires changing the way that the person selects, prepares, and consumes food—in other words, it requires changing not only what the person knows or thinks about food, but what a person *does*. Although this may seem obvious, somehow we often expect that providing information alone will automatically bring about change, and are surprised and frustrated when it does not. Why do so many clients say things like, "I knew what I was *supposed* to do instead of ordering a double cheeseburger and large fries, but I did it anyhow—I couldn't stop myself"? Providing generic information and authoritarian instruction rather than individualized counseling often simply adds to the client's feelings of lack of control and guilt, without facilitating health-promoting behavior. Change is difficult, and has costs as well as benefits;

sometimes the idea of change can seem threatening. Several models of health behavior and health decision-making provide useful insights.

## The Health Belief Model

According to Irwin Rosenstock's Health Belief Model, people arrive at a decision to make a change in their health-related behavior by doing a sort of cost-benefit analysis. They ask themselves (consciously or not) several questions: how bad will things get for me if I don't change? What will this change cost me, and what will get in my way? What will the benefit be if I do make this change? In the end, is the possible benefit worth the disruption of my habits, effort, time, dollars, loss of pleasure, etc. that making the change will cost me? As nutrition counselors, we must help the client answer "yes!" to that final question, by identifying benefits that are sufficiently motivating to that individual client while decreasing the difficulty and cost of making the needed changes.

This is the point where we need to become very much client-centered, because desirable benefits as well as costs are highly individual. It is essential that dietitians recognize that most people do not think the way that we do about food and nutrition. Most people are not fascinated (if not obsessed) enough with nutrition to devote their entire careers to it; most do not think primarily of the nutritional value of food or its impact on their health when choosing what to purchase and eat; most do not rank nutrition as the top priority in their lives. One could realistically describe a dietitian discussing food and nutrition with a typical lay person as cross-cultural communication. I have seen this clearly during the role-plays I assign to my Nutrition Counseling students—especially during the early weeks of the course, their discussion posts include many comments on how difficult it is to act the role of the client, pretending less knowledge of nutrition and adopting different priorities. For example: "I found it challenging to answer [as the assigned client] when role-playing. I found it difficult to make up foods. I really had to think of unhealthy options, think outside of what I eat … There were a few circumstances where I started to laugh because of what I was saying, when asking the counselor why I should eat vegetables and not fast food." The mental exercise of trying on the client's point of view and experience of the counseling enables the counselor to develop positive, acceptable suggestions and guidance. As another student put it: "I never thought about focusing on adding things to a client's life instead of removing things from their diet. It is definitely helpful though to play the role of the client and learn some about what it feels like to be in their place." The application of the Health Belief Model requires obtaining an accurate, empathetic picture of the client's life circumstances, baseline knowledge, goals and motivations.

## Stages of Change

An aspect of behavior change theory that is not directly addressed by the Health Belief Model is the client's readiness to consider, undertake, and maintain change.

The Transtheoretical Model, also known as Stages of Change (Prochaska and DiClimente 1983) is helpful in understanding where the client is in the change process. These stages can be described as follows:

- Pre-contemplation: The client has no awareness of a problem or a need to change. The counselor's role is to raise awareness.
- Contemplation: The client recognizes that there is a concern that might need attention. The counselor's role is to explore the issue with the client, providing information and helping the client to identify reasons to change.
- Planning: The client has decided that change is needed, but does not know what or how to change. The counselor's role is to help the client to identify realistic and desirable goals and explore ways to reach them.
- Action: The client is ready to implement specific changes. The counselor's role is to assist the client to strategize how to implement the actions, including identification of possible barriers and of resources that are needed.
- Maintenance: The client has successfully implemented changes and reached goals. The counselor's role is to support continuation of the changes with positive reinforcement, additional strategizing as new barriers arise, and identification of new approaches to prevent boredom or burnout.
- Relapse/recovery: The client has lost momentum, or has been confronted with new life circumstances that make the new behaviors more difficult. The counselor's role is to help the client to learn from these setbacks, develop new strategies, and resume the new behaviors without becoming mired in guilt and self-depreciation.

Conceptualizing the change process in this way can assist both counselor and client to focus on what is most helpful to the client, meeting the client where s/he is and building on from there. Not taking the client's stage of change readiness into account can easily result in a great deal of frustration (not to mention wasted time) for both parties. For example, imagine the effect of providing detailed instruction on carbohydrate counting (an action plan) to a client who has not clearly understood why this is relevant to his/her life and health (pre-contemplation stage). The likely result is a confused and possibly embarrassed or irritated client, and a dietitian who has completed a diet instruction but not effectively counseled the soon-to-be "non-compliant" client.

*Self-efficacy*

An additional piece of the puzzle of behavior change was conceptualized by Albert Bandura as self-efficacy. In brief, the idea is that people are far more likely to attempt and succeed at change if they believe that they are capable of executing the plan. This concept comes in to play mainly during the Action stage of change. Self-efficacy can be increased through ensuring that the client has a realistic plan with manageable goals, has a toolbox of strategies and resources for implementation that addresses potential barriers, and has confidence that the plan will work in his/her real world. Assisting the client in rehearsing and/or role-playing is often helpful, as is linking the client with outside resources such as support groups that offer successful role models and empathy.

*Client-centered Counseling*

Applying behavior change theory clearly requires the counselor to understand a great deal more about the client as an individual than the traditional diet instruction model anticipates. With the focus on the client's understanding, wants, needs, life circumstances, motivations and goals, the counselor must develop new patterns of communication.

## Counseling as Dialogue

The client-centered approach to counseling involves collaborative, two-way communication between counselor and client. Counseling is a goal-oriented conversation, requiring well-developed social communication skills.

*Setting the Scene*

The context in which nutrition counseling occurs has a great impact on its effectiveness. To promote a comfortable, respectful interchange, both the physical setting and the mindset of the counselor should be considered. A physical setting that allows for privacy, is pleasantly and neutrally decorated (e.g. pictures of beautiful food, yes; posters exhorting specific dietary regimens, no), and is set up to communicate collaboration rather than expert power (comfortable chairs of equal height, no desk forming a barrier between client and counselor) is ideal. In a setting where this is not possible, such as in a hospital room, at least the counselor can preserve the patient's dignity by asking whether it is a good time to talk, drawing curtains or closing doors, and sitting at bedside height rather than standing over the patient. The counselor should also mentally prepare to be fully present and focused on the client's needs. Taking a moment to "switch gears" from other activities, remove distractions, and commit to actively listen to the client is essential. Approaching the client with an attitude of curiosity, empathy, and respect sets the scene for a positive, collaborative dialogue. I recently heard a dietitian going into a meeting with a client say something like this: "His food records are a disaster! I don't know how people can eat like that—don't they know anything? I don't know what I'm going to do with him!" We need to remind ourselves that we are not the "food police," tasked with identifying nutrition crimes and doling out corrective sentences to our clients. We are there to identify client needs, help them to develop goals, and facilitate realistic and acceptable change.

*Starting the Conversation*

Nutrition counseling is both a professional and social interaction. As such, it should follow the social "niceties" that smooth our routine conversations, such as polite introductions of both parties and momentary "small talk" before launching into the business at hand. "Hello, are you Ms. Jones? I'm Laura Frank, the dietitian—how are you doing today?" This allows the counselor to signal that this is an adult-to-adult situation (and also verifies that the counselor is actually talking to the right person—not a bad idea in a hospital setting where patients often don't stay in the same place for very long). Briefly explaining the purpose of nutrition counseling is often helpful, to avoid or clear up misunderstandings. If the nutrition counselor presents him/herself as a supportive resource, available to answer questions and address client concerns about nutrition and health, this will reassure the client that s/he is not about to be judged or told how to eat, and will have input into the counseling session.

Many traditional nutrition counseling sessions move very quickly into taking a diet history or assessing "typical" eating patterns. However, for effective counseling to take place, the larger picture of the client's life situation has to be filled in first. I will never forget a patient encounter early in my internship. I was directed to see a young man diagnosed with kidney damage, and instruct him on a 500 mg. sodium diet. As I was about to begin the instruction, I asked him to tell me about his life situation. He told me that he was essentially homeless, living for short periods with a series of friends and relatives; he had little money, did not cook for himself and ate whatever food was available; and, he lived in a poor neighborhood where there was no supermarket where fresh produce or specialty low-sodium foods could be purchased. I looked down at the diet sheet and realized that it was completely irrelevant and useless to him. I don't remember what I recommended to him, but I learned that one-size-fits-all does not exist in the world of nutrition counseling.

Once the client can be seen in the context of his/her environment and lifestyle, it's time to move on to understanding eating patterns. Asking what the client "typically" eats is fraught with pitfalls, however. Many clients are aware of how they are "supposed" to be eating, and may be embarrassed or feel guilty describing their actual food choices; some will even "fake good" and report choices that they believe the dietitian will approve of while minimizing their "bad" choices. To avoid this defensiveness and take the discussion in a more positive direction, the counselor can start the conversation by asking, "What do you enjoy eating? What foods would you want to be sure to include in any eating plan we might decide on together?" This immediately lets the client know that her/his preferences are important, that s/he will not be asked to give up favorite foods, and that counseling will be collaborative. The counselor obtains a basis for building an acceptable eating plan based on honest disclosure of the client's actual choices, while the client is relieved of expecting to suffer in the name of good health.

*Interpersonal Communication Skills*

One fundamental difference between diet instruction and counseling is who is doing most of the talking. Ideally, the counselor should be primarily listening—specifically, listening actively, facilitating more client discussion by providing encouragement and feedback. An active listener:

- Mentally shifts roles from speaker to listener.
- Demonstrates attention both non-verbally (eye contact, body language, nodding and gestures) and verbally (encouragement, reflection), and pays attention to the client's non-verbal communication.
- Allows the client to finish speaking before starting to think of a response.
- Allows silence without "jumping in."
- Summarizes or paraphrases the client's statements, and checks for understanding and accuracy.
- Gives objective feedback regarding what s/he observes or hears.

More listening does mean less talking—particularly, less advice-giving. This can be very challenging for dietitians, who are trained to instruct. I assign a role-play to my beginning nutrition counseling students, many of whom are practicing nutrition professionals, and require that the counselor listen without giving advice for 15 minutes. As I walk around the room, listening to the role-plays, I hear advice slipping out within minutes. When these same students play the client's role, they are surprised at how off-putting it is to be given advice without first being known and understood—they talk about feeling shut down, cut off. Advice is helpful when the client wants it: "How can I tell how much sodium is in that bag of chips?" "Can you tell me which of the foods I like to eat are high in fiber?" If we're not sure whether a client wants advice, we can ask: "Would you like to go over a fast food restaurant menu with me to find out which foods are lower in calories?" Unsolicited dietary advice is no more welcomed by a client than we would welcome a colleague telling us what to select for lunch in the cafeteria at work.

When the nutrition counselor does speak during the counseling session, often it is to ask a question: to obtain information, check for understanding (the client's or the counselor's), request a clarification or amplification, ensure that the client's needs are being met. The way that a question is phrased can make a tremendous difference in the outcome. First, should the question be close-ended (yes or no answer) or open-ended? That depends on what is being sought. If it is a specific piece of information ("Have you ever been on insulin before to control your diabetes?"), then a close-ended question is appropriate. However, close-ended questions can also cut off important conversations or cause the counselor to make faulty assumptions. For example, the classic "Do you understand your diabetic diet?" If the answer is "yes," it could mean any of the following:

- Yes, I do understand it.
- No, but I'm too tired to talk about it now.
- No, but you'll think I'm stupid if I say no.
- I think somebody explained it to my daughter, so I don't need to know.
- I understand what you want me to do, but I don't understand why or what good it will do me.
- I understand what the diet sheet said, but I don't know how to actually choose foods to eat.

Far better to ask an open-ended question, such as "Please tell me what being on a diabetic diet means to you—how does it affect your food choices?" Open-ended questions allow the client to direct the response, and so elicit wider range of answers; they may reveal information or topics that were not thought of by counselor. To guide the direction of the session, the counselor can use a use series of increasingly focused "funneling" questions, going from general to more specific. One last point about wording of questions: using "what" or "how" questions is preferable to "why" questions, which can sound judgmental to a client. "Why do you think you weren't able to lose weight?" is likely to elicit answers like "I guess I just don't have any will power!", while "What has happened in the past when you tried to lose weight?" will probably produce a more objective and descriptive answer that will be a starting point for a useful discussion.

*Deciding on the Focus and Goals of Counseling*

If the counselor is to allow the client to guide the counseling process, several aspects of the client's activities, knowledge, beliefs, and preferences must be identified. What does the client want to change, and what does the client want/ need to NOT change? How does the client eat now, and what factors affect the client's choices? How ready is the client to make changes? What does the client know and believe about his/her health and about nutrition? What aspects of food/ eating are most important to the client? Here are some suggested questions to use in exploring these areas:

*General background/lifestyle:*

*Who lives with you?*

- How do you spend your time? (work, family, hobbies/interests)
- What foods do you enjoy eating?
- Who selects/buys/prepares the food that you eat?

*Setting the agenda/identifying the problem:*

- What are your health concerns?
- How are these concerns related to your food choices?
- What would you like to learn more about in these areas?
- Your doctor has referred you to me to discuss_____. What questions or concerns do you have about this?
- When you set the appointment, you said you were concerned about _____. Please tell me more about your concerns.

*Assessing motivation:*

- What reasons do you have for wanting to make some changes in the way that you eat?
- What barriers do you encounter in trying to eat the way that you would like to?
- How do you think you would benefit from making changes?
- How ready do you feel to start making changes today?

*Asking the client for direction:*

- What concern is most important to you right now? What would you like to work on first?
- What have you tried already, and how did that work for you?
- What do you most want to achieve?
- How would you like me to help you? What do you expect from me?
- What ideas do you have?

Initiating counseling as a dialogue sets up a process of shared responsibility for working toward mutually agreed-upon goals. The counselor acts as a professional guide, coach, resource, and support. To properly fulfill this role, the counselor should become educated as to the meanings of food in the client's life, beyond its impact on nutritional status and physical health. Food has tremendous emotional and psychological meanings that influence food habits and choices. Cultural, religious/spiritual, and socioeconomic influences also affect food choice. The remainder of this chapter will discuss issues of importance to the psychologically and culturally competent counselor.

## Emotional and Psychological Aspects of Eating

A diet order is like a medication prescription—here is the diagnosis, here are the foods to eat that will help to manage the condition. It's easy for the dietitian to get caught up in the simplicity of this equation, particularly now that we call what we do "medical nutrition therapy;" although this designation is necessary to promote reimbursement of our services, the term seems to minimize the complexity of the client's relationship with food. Further, changing one's diet often involves more than choosing to eat appropriate amounts of desirable foods: it also requires reducing or eliminating the intake of foods that might have value to the client for reasons beyond their nutrient content. Changing one's diet has much more "cost" than does choosing to follow a medication regimen.

It is, of course, not useful to view food as only a collection of substances that affect nutritional and health status. Most people choose to eat a food because it meets some other need: it tastes good, it is convenient, it is familiar, it is affordable, it has a social value—or, eating/not eating it serves an emotional or psychological purpose. When I teach in a weight management/lifestyle change program, in my other professional role as a psychologist, I hear repeatedly from participants that this aspect of eating is the hardest for them to deal with as they attempt to change their lifelong eating patterns. When we ask clients to change their food choices, we have to ensure that their emotional and psychological needs continue to be met, either with the new food choices or through non-food means, to support long-term success in maintaining these changes.

## Emotional Meanings of Foods

Once survival needs have been met, humans tend to select and eat foods for the sensory enjoyment and emotional satisfaction that they bring. Culinary traditions go back to the beginnings of human culture, and people who could afford it have always sought out new and stimulating eating experiences. Given this context, it seems obvious that nutrition counselors should take pleasure into account when guiding client food choices. I personally refuse to eat any food that I don't enjoy, no matter how "good for me" it may be; conversely, I make certain that foods that I deeply enjoy (chocolate, anyone?) find a place in my eating patterns. When I tell my clients this, I see visible signs of relief—I am letting them know that the "food police" have no place in our counseling process, and that healthy eating and eating pleasure can co-exist. Interestingly, when I tell my nutrition students the same thing, the reactions range from pleasant surprise to suspicion and disapproval. It sometimes seems that dietetics practice has a puritanical streak, expecting people to give up pleasure to be "good" eaters. A more positive approach would be to focus on encouraging clients to increasingly eat the healthier choices among their preferred foods, and to decrease the frequency and portion size of their less healthy preferences at the same time. There is considerable research supporting the effectiveness of this approach in supporting permanent changes in eating habits; concomitantly, other research shows that dietary plans that induce feelings of deprivation and guilt through restriction of food intake and choices often backfire. A review of literature published in the *Journal of the American Dietetic Association* (Polivy 1996) concluded: "Starvation and self-imposed dieting appear to result in eating binges once food is available and in psychological manifestations such as preoccupation with food and eating, increased emotional responsiveness and dysphoria, and distractibility. Caution is thus advisable in counseling clients to restrict their eating and diet to lose weight, as the negative sequelae may outweigh the benefits of restraining one's eating. Instead, healthful, balanced eating without specific food restrictions should be recommended as a long-term strategy to avoid the perils of restrictive dieting." A useful example of the positive approach to nutrition counseling is the Intuitive Eating approach of Evelyn Tribole and Elyse Resch, which seeks to re-train clients into a healthy relationship with food: eat when hungry, eat what you want to eat, stop eating when satisfied. In this approach, "satisfaction" refers mainly to a sensation that physical hunger has ceased, although it incorporates the idea of eating food that satisfies taste and emotion as well, allowing for occasional eating for enjoyment alone. Much of the counseling using this approach focuses on helping the client to tune in to the body, eat consciously, and recognize and respond to the client's personal signs of hunger and satisfaction. Applying this non-diet approach does require that the counselor and client be willing to trust in the client's inherent ability to make appropriate choices if given permission to do so. It also assumes that intuitive eating can be separated from the use of food to meet needs that are mainly emotional or psychological.

Unfortunately, for some clients food has become an emotional battleground. People may develop a love-hate relationship with food: constantly reading or thinking about food and eating, while fearing it as a source of dangerous temptation; struggling with control over their impulses, and filling with guilt and self-loathing when they lose that control. This can be seen in its most extreme form in individuals diagnosed with anorexia nervosa, who often enjoy cooking for and feeding others, while maintaining feelings of superior control by denying the food to themselves. Furthermore, many people use comforting or indulgent foods to soothe or self-medicate in the face of emotional distress. Although this happens for most people once in a while, when this pattern becomes a frequent and important coping mechanism it is beyond the scope of nutrition counselors to address these issues on their own. Such clients are best supported through collaboration with or referral to a competent psychotherapist. What nutrition counselors must realize, however, is that emotional or disordered eating is not primarily the result of the client not knowing about healthy food choices—in fact, many emotional or disordered eaters have a wealth of nutrition knowledge obtained from a variety of sources, although they may not have the educational background to judge its validity. Presenting nutrition education and convincing such a client to change eating habits without addressing the underlying emotional issues is unlikely to succeed, and will probably simply add to the client's feelings of guilt, inadequacy, and lack of control. Such clients tend to equate their weight and food choices with their personal self-worth, passing moral judgments on themselves when they fail to control their eating ("I have no will power, I make bad choices and can't control myself"). For such clients, the nutrition counselor should serve as a sounding board and support, helping the client to identify emotional eating patterns, to differentiate between problems that can be solved with food (i.e. hunger or a taste for a particular eating experience) vs. problems that require other forms of care, and to seek self-care and alternative coping mechanisms for non-food problems.

It is no accident that some people use food to self-medicate for emotional issues—there are well-documented links between food and mood. In fact, the *Diagnostic and Statistical Manual of Mental Disorders (DMS-IV)* lists persistent increase or decrease in appetite as a diagnostic criterion for major depression. I ask most clients to keep an emotional/environmental food diary, recording mood, level of hunger, social setting, associated activities, time of day, and thoughts along with the food selection. This often brings to light patterns of eating associated with one or more non-hunger triggers for eating that were previously unknown to the client. Many clients over or under eat when either distressed or happy, and react to emotional situations by seeking out "comfort" foods. We also often see patterns that reflect the influence of food composition on brain chemistry, particularly the serotonin-promoting effect of high carbohydrate foods—clients may seek out such foods when emotionally distressed. Awareness of such patterns of triggered eating is the first step toward learning new patterns of responses that avoid unhelpful food choices. The nutrition counselor can assist the client in developing alternative self-care responses to their emotional triggers, such as talking to a friend, taking a break, using relaxation techniques, and engaging in physical activity. Serious emotional issues, again, should be the domain of a qualified psychotherapist.

## Grief and Loss as Issues in Nutrition Counseling

When we speak of grief, we generally envision an individual mourning the loss of a loved one. Actually, grief can occur over many forms of loss; although we and they may not realize it, many of our clients are in the process of grieving when we meet with them to help them to improve their food choices and their health. For example, what has a newly-diagnosed diabetic lost that is worth grieving over? To begin with, s/he has lost the self-image of a healthy person, having to replace it with a self who has a chronic, life-threatening disease—often, this means facing one's mortality for the first time. Further, s/he has lost the freedom to choose food and even mealtimes freely without incurring dangerous consequences; may be facing further health losses due to diabetic complications; may be facing loss of income due to medical costs or work limitations; may even be facing relationship complications related to all of these changes. Life-changing illnesses can place our clients in the grieving process, which can bring about reactions and behaviors that seem irrational to the nutrition counselor. Elizabeth Kubler-Ross (1969) documented common stages of grief that can often be seen in our clients: denial ("I can't really have diabetes—the tests must be wrong. I don't need to change my diet"); anger ("This is just not fair! I shouldn't have to change the way I eat! I don't deserve this!"); bargaining ("OK, I'll never touch sweets again! I'll do whatever I have to do if it will cure me!"); depression ("What's the use in making all of these changes—the diabetes is going to kill me anyhow, just like my father.");

and finally acceptance ("Well, I guess I have to learn to live with it—what can I do to stay healthy?"). Understanding and recognizing this grieving process in our clients can help us to understand their responses and match our own responses appropriately. Trying to confront denial in a client who needs the denial as a temporary coping mechanism is unhelpful and frustrating to both parties—better to focus on general health promoting changes such as eating on a regular schedule than to teach carbohydrate counting for diabetic control. Understanding where anger is coming from can help the counselor to avoid taking it personally and permit a more positive focus. Recognizing bargaining can help the counselor to avoid false confidence and be prepared for the rebound when the changes turn out to be unmanageable. And, recognizing the signs of depression enables the counselor to avoid recommendations that the client is unable to cope with, while encouraging the client to seek emotional support—there are times when nutrition counseling has to go on the back burner, and our most helpful intervention is to refer to or consult with another professional to meet the client's needs. We can hope that our clients will reach a stage of acceptance, in which realistic change becomes possible and sustainable.

### Cultural, Religious/spiritual, and Socioeconomic Aspects of Food Choice

Cultural competency seems to be the newest buzzword in dietetics practice. Far from being the latest trendy point of view, culturally competency is coming to the forefront in a long-overdue recognition that food practices and beliefs have a tremendous impact on our clients' lives and on their response to our nutrition counseling efforts. It is essential that nutrition counselors educate themselves about multicultural foodways, communication patterns, and belief systems.

*Avoiding Cultural Paternalism*

Dietitians are trained to take on the role of expert in areas of food and nutrition—we promote ourselves as the "nutrition experts", and certainly do have the background to provide medical nutrition therapy from a Western Medicine scientific model. However, to effectively and sensitively provide nutrition counseling to all individuals, we need to acknowledge that there are multiple ways of seeing the world—some of which conflict with ours—that are not mutually exclusive. We may have to let go of the role of "expert", who seeks to replace ignorance with the truth as we see it, and enter into a process of mutual education with clients whose viewpoints and lifeways differ from ours.

All people see the world through a "cognitive filter," made up of our life experiences and the teachings of our families of origin and cultures. We learn ways of interpreting what we see and experience, and learn what is important or insignificant. Stop for a moment and ponder this: every one of us lives in a slightly (or greatly) different universe, even when sitting side-by-side. Nobody notices, sees, hears, interprets, values the world in exactly the same way as anybody else. Assuming that the counselor and client experience their world identically can lead to misunderstandings and conflicts. For example: think of the concept "corn." Most residents of the United States would envision corn on the cob, canned corn, maybe popcorn. Some residents might also think of regional or ethnic foods such as corn bread or corn tortillas. However, if you were a member of many of the First Nations peoples (Native Americans), you might think of many different words for "corn," representing seeds, green corn, corn in the field, dried corn, a corn deity, corn rituals, the cycle of planting and harvest, and more. "Corn" may be so central to your life and culture that one word would not suffice to convey the complexity of the concept. Imagine the disconnect between a nutrition counselor who regards corn as a food with a typical nutrient profile and glycemic index, and a client for whom corn is at the center of much or his/her culture, when the nutrition counselor proceeds to teach the diabetic client about carbohydrate counting. There is going to be a great need for mutual education and a search for common ground before this encounter can be productive. It becomes essential that the nutrition counselor remain open to learning from the client about his/her experiences and views. Asking the client to teach does more than increase the counselor's knowledge: it empowers the client, and conveys value and respect for the client and his/her culture.

There are many ways of seeing the individual, his/her place in the world, the nature and process of health and disease, and the role of food in all of these. Dietetic "culture," based in Western scientific medicine, sees disease as caused by physical or metabolic errors, disease organisms, and deficiencies or excesses of biochemical substances; the role of food in health is to provide the nutrients that support healthy physiologic function, and to *not* provide substances that might contribute to disease processes. Many Asian and Latino cultures see disease as caused by imbalances of categories of metabolic and spiritual energies, for instance, "yin and yang" or "hot and cold;" the role of food in health is to not only provide the material substances needed for health, but to maintain or correct the balance of energies within the physical and spiritual self. This can sometimes cause conflict where these ideas clash. For example, a large teaching hospital located in the "Chinatown" section of Philadelphia saw the nursing staff in its maternity area in frequent conflict with its traditional Asian clients: patients were refusing to eat the meals being delivered to the floors, families seemed distressed and avoidant of the staff, and staff had to contend with family members bringing in food to the patients. In addition to the language barrier, the patients' culture of origin valued privacy, modesty, and avoidance of confrontation, so it took considerable work with community members before the hospital uncovered the source of the problem. It turned out that the foods designated as medically appropriate by the staff, such as jello, were viewed by the patients' culture as having the wrong balance of food energies to promote the health of pregnant and postpartum women—in their belief system, the hospital was providing food that would sap the strength of the women and slow their recovery and ability to breastfeed. Once the hospital changed the food selections, the community became a major source of patient referrals—patients and health care providers had found a common ground and mutual respect for each other's system of food–based health care.

In working in cross-cultural situations, identifying "gatekeepers" (opinion leaders, authority figures, elders) who have the respect of the community and can act as a bridge between the community members and the health care system is essential. Once this connection is established and the provider has been accepted as credible, approaches can be worked out to create a positive cultural and social context for counseling. It is also helpful for the counselor to become aware of his/her own cultural food traditions and perceptions, and be willing to share with clients in a mutual exchange. Most important is to avoid setting up a choice between your way and the client's—given a choice between generations of tradition and the advice of a stranger, the outcome is predictable!

*Socioeconomic Considerations in Counseling*

In considering the cultural context of the client, there are many other issues that could be addressed. Sensitivity to socioeconomic constraints is a necessary precursor to providing nutrition counseling that is usable to the client. Two significant issues are food access and literacy.

Identifying issues related to food access requires going beneath the surface of the client's reported dietary intake. Simply educating the client as to what foods would be healthful to choose is not only irrelevant, but embarrassing to the client, if those healthful foods are out of the client's reach. Many clients do live in "food deserts:" surrounded by fast food outlets and convenience stores with no food market for miles around. Fast food is cheap, requires no preparation (useful if one doesn't have a working stove or refrigerator, or even electricity), is filling, and easy to find—none of that applies to whole grains, fruits and vegetables. Clients may not be forthcoming about food access barriers, so it is the counselor's responsibility to become educated about the client's community and life situation. Developing a resource list that can connect clients with free or affordable healthy foods would be of value to all clients, regardless of economic status; such resources can be offered to any client without putting the client on the spot to admit to having limited personal resources.

Food access also includes knowing what to do with the food once the client has it. Providing this information in a culturally appropriate manner may require education of the counselor as well as the client. For example, one of my students recognized that the largely Hispanic clientele of her WIC office were apparently not using the fruit and vegetable vouchers that they had begun to receive. She realized that the WIC nutritionists, who were not Hispanic, were not sufficiently knowledgeable of the food ways of their clients to advise them how to use the fruits and vegetables that they could purchase locally with the vouchers. The student's thesis research thus involved developing an education program for the

WIC nutritionists on the foodways of their clients, enabling them in turn provide culturally relevant counseling to the clients.

Access to nutrition knowledge depends on the information being provided in an accessible format. The counselor should not assume that the client is able to read written information, or that the client understands verbal information that uses technical wording or is not in the client's native language. I still recall as an intern observing a nutrition counseling session in a Southern California county hospital between a Midwestern Caucasian dietitian and her Hispanic pregnant client. "Tu tienes comer comidas con calcio, por su nino. Tienes que beber mas leche" (as far as she knew, she was saying, "You have to eat foods with calcium for your baby. You have to drink more milk"). While the dietitian seemed satisfied that she had completed a successful counseling session, the young woman seemed perplexed. And why not? Setting aside issues of grammar and pronunciation, there was no explanation in any language as to what calcium was, what it had to do with milk, or why her unborn baby needed her to drink milk when many adult members of her community were lactase-deficient and avoided drinking milk for cultural reasons.

Culturally competent counseling is a mutually enriching interchange between counselor and client. Learning about ourselves and our own food history, then further enriching this experience through learning about our clients' foodways and lives, can be an unending and fascinating journey. Food is a window into the culture, traditions, beliefs and history of a civilization, and we all share a need to eat and be nourished.

## Reconnecting with Food

Over the past century, people in the United Stated, have gradually, then rapidly, become distanced from the sources of our food. Popular food writers such as Michael Pollan (*The Omnivore's Dilemma*), Marion Nestle (*Food Politics*), and Eric Schlosser (*Fast Food Nation*) have documented the rise of fast food, agribusiness, food marketing, and food processing, while the nutrition literature has documented the parallel rise in obesity and many chronic diseases. From an agricultural society where many people raised their own food, we have developed into a society where many people have never met a farmer, seen a living farm animal or picked a tomato, and food is recognized by the package it comes in from the store. Food is often consumed rather than enjoyed, with convenience and speed eclipsing freshness, taste, and aesthetic enjoyment. "Value" is marketed as "more for your money," with "more" meaning quantity rather than quality. Although this may seem to be outside of the realm of nutrition counseling, it is actually the context in which nutrition counseling takes place.

When people make the effort to consciously choose good food and take the time to enjoy it, they nourish themselves on multiple levels: physically, socially, culturally, emotionally, perhaps even spiritually. I like the definition of "quality food" as proposed by Carlo Petrini, the founder of the Slow Food Movement: food should be "good, clean and fair"—healthful and delicious, sustainably produced, and equitably grown and distributed. It is possible to develop access to good food for individuals of all socioeconomic levels—even "food deserts" can grow gardens and host farmer's markets when people are educated and empowered, as the Food Trust and similar organizations are demonstrating with their community partnerships. I believe that an important goal of nutrition counseling and food-related education should be to promote an understanding of the many roles of food in people's lives. Beyond that, nutrition professionals and all who are passionate about food and nutrition should advocate for all people to have the option of accessing and enjoying good food that is appropriate to them as individuals, families, communities, and cultures. The act of conscious eating connects all of us to the earth that supplies our food, the people who grow and prepare it, the cultures that inspire it, and our personal and communal histories. Growing, cooking, sharing, and enjoying food has linked people together and sustained us throughout human history. Nutrition, nourishment, nurturing—all are aspects of healthful eating.

### References

Kübler-Ross, E. 1969. *On Death and Dying*. New York: Routledge.

Nestle, Marion. 2002. *Food Politics: How the Food Industry Influences Nutrition and Health*. Berkeley: University of California Press.

Petrini, C. 2007. *Slow Food Nation: Why Our Food Should Be Good, Clean, And Fair*, New York: Rizzoli Ex Libris.

Polivy, J. 1996. Psychological Consequences of Food Restriction, *J Am Diet Assoc*. 1996; 96: 589-592.

Pollan, M. 2006. *The Omnivore's Dilemma: A Natural History of Four Meals*. New York: The Penguin Press.

Prochaska, J.O. and DiClemente, C.C. 1983. Stages and processes of self-change of smoking: Toward an integrative model of change. *Journal of Consulting and Clinical Psychology*, Vol 51(3), June 1983, 390-395.

Schlosser, Eric. 2001. *Fast Food Nation*. New York: Houghton Mifflin Co.

Tribole, E. and Resch, E. 2003. *Intuitive Eating, A Revolutionary Program that Works*, St. Martin's Griffin. 2nd edition.

# Thematic Tabs for Chapter 7

*Access*

*Body*

*Colonial*

*Emotion*

*Nature*

*Race*

*Women*

Chapter 7

# Doing Veganism Differently: Racialized Trauma and the Personal Journey Towards Vegan Healing

A. Breeze Harper

**Editors' Note**: This chapter links to the themes of Access, Body, Colonial, Emotion, Nature, Race, and Women. Harper powerfully explores the experience of racial trauma and normative whiteness and the interconnection of these experiences with a body-food relationship high in processed foods and leading to bodily ill-health. She leads us on her own journey of bodily decolonization, which centered on a plant-based diet and connected her with many other African diasporic females who are doing the same – that is, women whose vegan practices are firmly rooted in a de-colonial body politics. These women became part of her *Sistah Vegan* project, an online community and published anthology.

I have written this chapter with the intention of reflecting on my personal journey towards becoming a vegan, and eventually creating *The Sistah Vegan Project* and *Sistah Vegan* book. I will be sharing how and why my personal experience as a black racialized female subject, in the USA, produced a relationship with food, healing, and veganism that is distinctly different from the white and upper-middle class norm that undergird the "common sense" rhetoric of USA mainstream alternative food movement. Such racialized distinctions are important for those interested in approaching nutrition and health education in the USA. Despite the Civil Rights Act, we in the USA continue to be negatively affected by racial, class, and gendered configurations of power, privilege, exploitation, and discrimination. This chapter shares how such configurations have even affected one's conception of food, the practice of veganism, and the act of decolonization. For people like myself (black and vegan), veganism becomes a way in which we seek to decolonize the negative effects of colonialism on our bodies and minds. I first start with vegan eco-chef, Bryant Terry.

On March 3, 2009, the book release party for *Vegan Soul Kitchen* took place in San Francisco at the Museum of the African Diaspora. The author, Bryant Terry, was being featured as part of "Chefs of the African Diaspora" series. Terry is unique within the genre of vegan cookbooks, as he is African American and male; USA authored vegan cookbooks are basically the domain of white. When Terry took the stage to introduce himself and his new book, the wall in back of him had a projected image of yellow and pink watermelons. He explained to the audience that the projector was supposed to be showing a slideshow, however, it remained stuck on that image of the watermelon due to mechanical failure. He conveyed to us how it was appropriate that it would be stuck on that particular image. Terry explained that he had not eaten a slice of watermelon until he was seventeen years old. His parents were fearful of consuming watermelon because of the negative stereotypes associated with black people. What made Terry's comment even more interesting was that my husband, a white male born and raised in Munich, Germany, didn't understand what Terry was talking about. Of course this makes sense, as there is an entire racialized narrative around the inferiority of Black people and their weakness for both watermelon and chicken within the USA (Williams-Forson 2006; Witt 2004).

Despite my husband's lack of awareness to the racialized history of watermelon in the USA, the audience of mostly brown and black people made bodily gestures and sounds that indicated that they knew exactly what Bryant was talking about and how he had felt. This spanned beyond just being about watermelon; as the watermelon merely stood as a symbol of how racialization in the USA can affect one's consciousness, as well as what they find pleasurable or painful to consume. Terry's confession about his history with the watermelon was both saddening yet inspiring to me. It led me to ask the question, *How has being racialized-sexualized as a black American man, born and raised in Tennessee, affected Bryant Terry's construction of the vegan recipe book Soul Food?*

The intention of Terry's *Vegan Soul Kitchen* cookbook is to "reclaim" soul food in a way that is positive; in a way that means black and brown people should be able to consume without the consequences of being branded with racist stereotypes that accompany this cuisine (Williams-Forson 2006; Witt 2004). In addition, Terry describes his recipes as bringing soul food back to its "healthier" state before the industrialized US food industry took over most of America's dinner plates. Terry's book focuses on the wholesome goodness of soul food. But, importantly, Terry's soul food is cooked without ingredients that he thinks perpetuate the nutritional-based health disparities that continue to rise in black and brown communities throughout the US. Such ingredients include refined and bleached flour, table salt, sugar and high saturated fat animal products (Terry 2010) that can be considered as being part of a "colonized" diet.

After I left the book release celebration, I could not stop thinking about how my mother would not let me engage in certain activities that were markedly "stereotypically black." My parents raised my brother and I in Lebanon, Connecticut, a rural New England town of which the population is about 98% white. I remember wanting to learn how to tap dance but my mother had absolutely prohibited me from doing so. It was absolutely too painful for my mother to imagine seeing her own daughter tap-dancing while white people looked on. Would they interpret the meaning of the dance through their own white ontological lens and see me as buffoon? Would my performance be too much of a minstrel show? Mom thought that I would become an object for white folk's fantasy world of how black bodies should perform, for the sake of white entertainment. Such objectification of the black racialized body is one of the core psychoanalytical issues that Algerian theorist, Frantz Fanon, focused on in his own work against the racist colonial project. Terry's parents most likely had the same fear for their son that my parents had for me. Both watermelon and tap dancing are symbolic of acceptable and unacceptable performances of blackness. Though dance is an obvious example of actual performance, food can be more subtle in how they symbolize racial formation and social hierarchies of power and privilege.

These anecdotes have led me to some important, and difficult, questions. How does one simultaneously understand the impact of their personal racialized trauma, their performance of blackness (such as soul food consumption), and be *self-reflexive* about its impact on their construction of vegan food activism? I see texts like *Vegan Soul Kitchen* as about creating *black* healing across several spatial scales: black minds, black bodies, black kitchens, and black communities. Such a healing project is challenging and necessary; Terry's food activism takes the trauma of racism and creates healing solutions through vegan health activism. He is not the only African American vegan engaged in such activism. Myself, and vegans such as Queen Afua (briefly explored later) are vegans who engage with intersections of racial trauma, health disparities, and food justice in our race-conscious approach to veganism. How can trauma be a platform of learning and activism?

In her book, *An Archive of Feelings*, Ann Cvetkovich's focus is on trauma culture and being "queer" in the USA. She writes,

> A significant body of work within American studies has recently mounted a critique of U.S. culture by describing it as trauma culture. Wendy Brown speaks about identity politics as a politics of ressentiment in which claims on the state are made by individuals and groups who constitute themselves as injured victims whose grievances demand redress ... Lauren Berlant develops the notion of an 'intimate public sphere,' the result of a process whereby a 'citizen is defined as a person traumatized by some aspect of life in the United States.' (Cvetkovich 2003, 15)

Cvetkovich's analysis of trauma is very interdisciplinary, and even though she acknowledges that trauma studies have been traditionally rooted in psychology, she seeks to "demedicalize" and "depathologize" its usage by turning to "feminist theory, critical race theory, Marxist cultural theory, and queer theory" (Cvetkovich 2003, 12). What she attempts to do is to bring the subject of trauma into the public sphere while trying not to pathologize people who have been traumatized. I think one of the most important questions Cvetkovich's book asks is, *What public cultures are created around traumatic events?* Such an emphasis helps to shift trauma as a medicalized concept, found in the clinical text Diagnostic and Statistical Manual of Mental Disorders, to a social context that is effective on body politics and group identity formation. So, when Bryant Terry presented at MoAD (Museum of the African Diaspora), it was part of a "reclaiming" event that wanted to publicly celebrate soul food. This celebration was two-fold. First it wanted to celebrate the beauty and genius of soul food cuisine. Secondly, the celebration and Terry's talk recognizes the emotional pain that many black and brown people may have had to deal with when deciding if they should consume soul food in certain public spaces. In her book, *Building Houses out of Chicken Legs: Black Women, Food, and Power*, Williams-Forson (2006) writes of how a significant number of the black middle-class women wanted to teach working-class black women a more Euro-centric 'gourmet' way of cooking food for their families. Furthermore, these black middle class women were determined to teach these women that racial uplift couldn't not be properly achieved if black people continued eating 'stereotypically black' food in public spaces, such as fried chicken (Williams-Forson 2006). It is clear to me that the MoAD's "Chefs of the African Diaspora" seeks to find a more empowering aspect to soul food that has gotten lost, particularly within the recent context of mainstream nutritional professional pathologizing soul food as the "root" cause of black racial health disparities. Instead, this artistic exhibition shows how soul food is an act of agency and sublimation for black bodied people in white colonial spaces; that soul food is not static, but can be transformed in a way that addresses racialized health disparities without judging the entire culture of soul food as pathological and unredeemable.

In returning to my own experiences, I wonder if I have come to know my position as a "black" subject because of both racialized trauma and the agency and sublimation that I have finally been afforded to heal these experiences. For example, I am creator of the Sistah Vegan Project, a book anthology and online community of African Diasporic females that practice decolonization of their bodies and minds through plant-based diets. Lantern Books published this seminal book about black female vegan experiences in March 2010, titled *Sistah Vegan: Black Female Vegans Speak on Food, Identity, Health and Society*. Though I had been writing and editing the project for nearly four years, it wasn't until I was publicly acknowledged that my writing entered a "third space" of healing and reclaiming through what Kelly Oliver calls sublimation:

> Sublimation is the linchpin of what I propose as psychoanalytic social theory, for it is sublimation that makes idealization possible. And without idealization we can neither conceptualize our experience nor set goals for ourselves; without the ability to idealize, we cannot imagine our situation otherwise, that is, without idealization we cannot resist domination. Sublimation and idealization are necessary not only for psychic life but also for transformative and restorative resistance to oppression ... It is through the social relationality of bodies that sublimation is possible. But in an oppressive culture that abjects, excludes, or marginalizes certain groups or types of bodies, sublimation and idealization can become the privilege of dominant groups. (Oliver 2004, xx)

I had always written privately about my racialized, embodied perceptions of the world. However, it wasn't until I finally found a press to publish *Sistah Vegan* that an intense feeling of healing and reclaiming of my embodied experience as a black female in the United States overcame me. Finally, I could express my experiences of practicing a vegan philosophy as a black female in a white middle-class dominated vegan USA – a world where the politics of whiteness, racial and class privilege, and covert racism are rarely, if ever, brought to light (Adams 1994; Torres 2008). For many non-white racialized people in the United States (vegan or not-), this silence is traumatizing, if not emotionally immobilizing (Ahmed 2007; Leary 2005; Oliver 2004).

However, I have also begun to reflect on why I should feel emotionally "better" when the white public mainstream allows me to visibly sublimate my own racialized experiences with reality. Am I caught in a paradoxical relationship? Do I need White America's approval of *Sistah Vegan* to feel like a human being? These are very deeply personal and complex questions that I feel I can no longer relegate to the physical and psychic spaces of the private. I have been inspired by Bryant Terry's public reflections on his and his parents' black embodied experiences with racial baggage over something that should be pleasurable (eating watermelon) yet is symbolically shameful and traumatizing. Terry's *Vegan Soul Kitchen* is an example of what veganism and soul food look like, from the perspective of a black male who wants decolonize, yet simultaneously celebrate, soul food cuisine. Terry's narrative of growing up has inspired me to reflect on how my own racial trauma influenced my food justice, decolonization, and veganism. Here is my story.

I started the Sistah Vegan Project in 2005, because it has been increasingly clear that even though a properly planned plant-based diet can combat the nutrition related diseases we see within the Black community in the USA, a challenge I have noted is that the mainstream vegan literature's intended audience is, by default, white and middle class – a demographic whose collective consciousness has been wholly influenced by understanding their relationship to consumption and their bodies, through spaces of race and class privilege. The popular online bookstore, Amazon.com, lists their top selling 100 'vegan' books. With the exception of Alicia Simpson's *Quick and Easy Vegan Comfort Food*, the other 99 titles are not authored by African Americans; furthermore, none of these vegan books include a section that speaks about the impact that racist trauma (or even normative whiteness) has had on how they develop their socio-spatial vegan epistemologies.

Racialized places and spaces are at the foundation of how most of us develop our socio-spatial epistemological grid – that is the ways in which we make sense of the social and geographic world around us, based on spatial relations and what our bodies mean in certain places (Dwyer and Jones III 2000; Lee and Lutz, 2005; McKittrick and Woods 2007). Hence, we come to make sense of the world in ways that are deeply racialized. For example, collectively, low-income urban Black Americans in the USA *know* that a holistic plant-based diet is nearly impossible to achieve; simultaneously, the collectivity of white middle-class urban people in the USA *know* that a holistic plant-based diet is easy to achieve (see Dubowitz et al. 2008). If you are the former, your relationship with healthier food options is influenced by environmental racism, lack of access to public transportation to get to healthier food sources, and the placement of fast food and liquor store chains in closer proximity to you than an affordable produce center (see Baker et al. 2006; see Cohen et al. 2010). If you are the latter, a combination of white and class privilege have socially and physically placed you in a location of environmental privilege, easy access to transportation to get to healthier options, and the placement of holistic health-oriented food locations (CSAs, Farmer's Markets, produce grocers, etc) that exist in your town (see Wilson et al. 2008). Such racialization of food access is, of course, linked to class, and class is certainly linked to the question of who gets to live in healthier environments. However, access to "healthier" food isn't just about class. Despite even having high socio-economic status, middle class Blacks in the US have significantly less access to healthier nutrition sources then whites from the same class (Freeman 2007; Randall 2006).

The way vegans in the USA think about a moral food system cannot be separated from the places and spaces within which they have been racialized; hence, these epistemologies are racialized and vegan epistemologies are no exception. The reason that 'race-conscious' books such as *Sistah Vegan* and *Vegan Soul Kitchen* exist, is because they directly address issues of racialized food access and knowledge. These topics are elided within bestselling vegan oriented books such as *Skinny Bitch*, *Quantum Wellness*, and *The Kind Diet*, which take a post-racial and white normative approach to food, health, and nutrition in general (Harper 2011).

The Sistah Vegan project reexamines veganism as an alternative, food ways movement, as well as a personal health choice through a Black feminist, critical race, and decolonial analysis. The Sistah Vegan project started as an online community, to start formulating answers to the following questions we have seen within veganism:

- How are Black female vegans using whole foods veganism to decolonize their bodies and engage in health activism that resists institutionalized racism and neocolonialism?
- If a majority of Black people have had negative experiences with "whiteness as the norm," and they have come to believe that veganism is a "white thing" that is disconnected from anti-racist activism, how can sistah vegans and allies present a veg model as a tool that simultaneously resists (a) institutionalized racism, (b) environmental degradation, and (c) high rates of health diseases plaguing the Black community?

However, how I got to this point in my consciousness must be located somewhere in the 1990s. These questions didn't come to me overnight, of course. It was my exposure to a canon of black feminist scholarship as a college student that would eventually help me develop a critical race and black feminism framing of food, power, and race in the USA, one decade after I had graduated.

I think it is important for me to use my own personal experiences with racism, as well as coming from a working class background, to illustrate how manifestations of race (trauma from racism, racialization, anti-racism) have (1) affected my transition into a plant-based diet as a healing system against such traumas; (2) influenced me to create the *Sistah Vegan* anthology, and (3) led me to apply a critical race feminist analysis of the US vegan movement as my dissertation project.

First, I was *not* born and raised in an environment in which my parents did not have access to what I needed to eat a healthier diet. So, my experience contrasts the overwhelming scholarly focus on 'access' as a cure-all for healthy eating; that is, the presumption that all Black people need to be healthier is access to healthier foods as well as community gardens to grow these foods and/or natural food coops. The literature has also suggested that it is the phenomenon of "food deserts" in the urban environment that makes it difficult for Black people to "eat better," implying that rural blacks don't necessarily have this challenge. Such literature thus inadvertently advances over-generalizations about the Black population. I was raised in an all white rural New England working class town. My parents took a mortgage out to afford a house on 2.5 acres of land that my father was obsessed with turning into "edible landscaping." We had an impressively diverse garden and orchard. My father taught me how to grow my own food; our family also traveled to the town's natural food store, once a week, so my father could teach my twin brother and I how to eat "healthier" foods. In theory, I had everything that I needed, to pursue a healthier diet, but simply could not motivate myself to eat healthier. Something kept occurring, over and over again in my life that often ended up being more powerful than the roots of the peach, apple, and walnut trees in our family's yard: racist ideologies, circumstances, and spaces that were nearly impossible for me to avoid consuming. Racism was so deeply traumatizing to me, that I didn't realize until over 15 year later, I had dealt with this pain through comfort eating of what could be considered bad or junk foods with highly processed ingredients such as bleached flour, corn syrup, refined sugar that are known to cause health problems (Barnard 1993) and are deeply colonialist.

On the first day of the 7th grade, someone said loudly, "Look at that skinny little nigger. Run nigger run." I was absolutely terrified. During that day, I remember overdosing on Smarties candies as a way to deal with what had happened, and as I look back, I realize that I did this repeatedly to deal with the ongoing racisms thrown at me. When I went to Dartmouth College with my twin brother in the fall of 1994, I was no longer under my parent's roof. I could literally eat anything I wanted to, to deal with the dynastic class privileged ruling elite attitude that permeated the entire campus.

One day, during the winter of 1995 my twin brother came into my dorm room, my two other roommates there. He happily told me, "Guess what? Dartmouth Office of Financial Aid gave us more money so I can have a better meal plan and mom and dad don't have to struggle so much," and then he left back to his dorm room. My roommate, Liz, a woman from California (who was white and upper-middle class) confidently and uncompassionately said, "Dude, I bet they only gave your brother more financial aid because he's Black!" I was shocked, stunned, angry, but too scared to say anything back to her (I realize now that I was following that internalized colonialism and cultural conditioning to not 'upset' white people when they speak with ignorance about black folk). Instead, I remember feeling the hunger to speak up and yell at her being replaced with the hunger to go to the campus cafe and buy a cheeseburger and cheese fries, despite me knowing full well that I am severely lactose intolerant and that I would be awake all night, sick from ingesting these foods. I know now that whenever the stresses of racism, classism, and sexism hit me during my earlier years, grabbing for Chicken McNuggets, Dr. Pepper soda, and Hershey's milk chocolate bars were a BandAid for my hurt emotions. Unfortunately, these products did not nurture my body in a way that can combat stress and resist physical ailments that manifest from ongoing racist tensions. Despite Dartmouth College being a very green and health foods oriented campus, I sought solace in the foods that many of my peers looked down upon with disgust.

I was so engaged with surviving through the weekly assaults of racism, classism, sexism, and white privileged entitlement that I felt "green" living, "green" eating, etc., were not applicable to my situation. (It wasn't really until my sophomore year in college that I had the language – or rather an entire canon of black feminist thought – to start deciphering and analyzing my past experiences and being able to name these verbal assaults as 'racist' or 'white privileged.') At Dartmouth College, students were expected to embrace the white middle-classed experience of "green" activism, all while the racism that pervaded the campus *never* entered into the conversation of "greener" diets. I found it contradictory that many of my white class privileged peers would tell me I should become vegetarian or vegan *because of animal rights*, or that eating a burger is "unethical" because of the land wasted for grazing cattle. I was being told this by peers who seemed more interested in non-human animals than considering what it meant for they themselves to never talk to me about their very own racial and socio-economic privilege. Not once did they want to reflect on what these privileges meant. For example, such privileges meant that they have had easier access to healthier food choices, healthier job opportunities, and healthier residential living choices (Gottlieb and Josh 2010; Liu and Apollon 2011; Randall 2006). In retrospect, I wish someone would have approached me and asked, "Breeze, being at Dartmouth as a black and working class kid is tough. I notice that you're dealing with racism, sexism, and classism here at Dartmouth. You're also eating crappy food as a comfort device to deal with these stresses. Consider a plant-based diet as a way to decolonize your body, you know, from the ongoing stresses of racism and elite classist energies here" I had read bell hooks' *black Looks: Race and Representation* for my introductory class to Women's Studies. It was through her that I learned of the concept of decolonization, however, my thinking was still too immature and single-issue to apply decolonization and anti-racism to food and nutrition. In her book *Breaking Bread*, I read this below in 1995:

We deal with racist assault by buying something to compensate for feelings of wounded pride and self-esteem...We also don't talk enough about food addiction alone or as a prelude to drug and alcohol addiction. Yet, many of us are growing up daily in homes where food is another way in which we comfort ourselves.

> Think about the proliferation of junk food in Black communities. You can go to any Black community and see Black folks of all ages gobbling up junk food morning, noon, and night. I would like to suggest that the feeling those kids are getting when they're stuffing Big Macs, Pepsi, and barbecue potato chips down their throats is similar to the ecstatic, blissful moment of the narcotics addict. (hooks 1993)

I remember at the time that I didn't quite 'get' what she was saying. I interpreted it that "of course black folk eat foods that kills us. We're stressed as hell about the racist bullshit assaults on our lives everyday. Good, hooks named what I am going through, so I guess it's okay that I'm eating my Hostess cherry pies." It didn't occur to me that I should stop eating that way and seek an alternative. Perhaps it didn't occur to me because it was just a small paragraph in a book that wasn't even about food or nutritional decolonization. Perhaps it didn't occur to me because I was still surrounded by rhetoric of "going green" and "eating healthy" that was repulsively post-racial and dismissive of the lived realities of those who weren't benefiting from the value system of upper-middle class whiteness.

Unfortunately, the healthy-eating message always came from a vantage point in which anti-racist and anti-classist activisms were not part of the rationale for transitioning into a plant-based or vegan diet. Listening to them, I knew something was amiss because my peers of race and class privilege spoke of food and ethics from a vantage point foreign to me. It was quite obvious they had not spent any time in the inner city of Hartford, Connecticut, where my grandparents lived and my family visited nearly every weekend. I remember there always being liquor stores or convenient stores with bars on the windows, dilapidated buildings, cement everywhere, and passing by a toxic smelling dump to get to grandma and grandpa's home. It would take me another ten years to realize that my beloved grandparents and their community were recipients of what Bullard calls environmental racism (Bullard 1993).

The narrative I am sharing above is not a challenge that I alone have encountered, but a challenge that a significant number of other black females on the Sistah Vegan project encountered as well: Though the benefits of a well-planned, plant-based diet are exceptional, time and time again, the language used by the mainstream to convey this message frequently ignores the lived realities of racism and its effects on the bodies, minds, and souls of people of color. However, this makes sense because the mainstream spin on veganism and vegetarianism is from the vantage point of the middle-class and white racialized embodied experience in the US (Adams 1994; Torres 2008); their consciousness has been shaped by such racial and class locations, which informs their sense of food, ethics, and justice.

Here in the global West, our epistemological grids around issues of morality and justice, including food activism, are directly influenced by our visceral experiences with race (i.e. our lived experience of practices of racialization, racism, racial privilege, and normative whiteness). The collective rhetoric of my white pro-local, pro-Organic, and/or pro-vegetarian peers at Dartmouth College was stemming from what Farr (2004) calls a "white racialized consciousness." Defined by African American philosopher Dr. Arnold Farr, racialized consciousness:

> replace[s] racism as the traditional operative term in discourses on race. The concept of racialized consciousness will help us examine the ways in which consciousness is shaped in terms of racist social structures ... 'Racialized consciousness' is a term that will help us understand why even the well-intentioned white liberal who has participated in the struggle against racism may perpetuate a form of racism unintentionally. (Farr 2004, 144-145)

Racialized consciousness is not necessarily the same as being *racist*. Instead, the term operates as a way to better understand those white people who do not fully understand that they are engaging in covert acts of whiteness/white privilege racism when they believe they are sincerely engaging in activism like food justice. Having lived in a nation in which their white epistemologies and ontologies made up the status quo, my white class privileged peers at Dartmouth College were unaware of how their white racialized consciousness does not reflect the reality of those who do not exist in white privileged spaces of inclusion. Collectively, Black female US Americans such as myself, have different understandings of consumption – philosophically, materially, and psychically (Williams-Forson 2006).

After four years of Dartmouth College and hearing endless white class privileged moral logics of "green" justice: go veg for Animal Rights (versus human rights), Sierra Club style environmentalism (vs. anti-racist Ella Baker type of environmental justice), I graduated from Dartmouth College as a fast-food omnivore. I had been unimpressed and unmoved by the moral logics of the status quo. At this point I still did not understand what organic food, vegetarianism or veganism, or food co-op housing had to do with resolving the trauma and pain I felt from the dynastic, white class privileged, sexist, homophobic, and isolating place that was my experience at Dartmouth College in the 1990s. It would take more bodily trauma – a diagnosis of a tumor – to help me to understand how veganism, organic food, and holistic health can be rooted in battling the bodily and emotional traumas of racism for Black women such as myself.

I graduated from Dartmouth College in 1998, lived in New Jersey for a year, and then moved to Boston in 2000. In 2002, I was diagnosed with a fibroid tumor in my uterus. I immediately called my dad after leaving the gynecologists' office to tell him, "I'm so upset! I have a fibroid tumor just like mom!" My mother was in her early thirties when she had a hysterectomy because of her fibroid. A few days later, my father emailed me and suggested, "Well, why don't you see what our people used as healing herbs for women's issues back in the day? What did we use for women's health?" Pursuing that question was intensely challenging. My gynecologist had recommended that I take birth control pills or consider a uterine myomectomy if the tumor starting causing major problems. I decided I didn't want to choose either of those options, and even though I was exercising every day, had lost weight and was down to a BMI of 21, I did not fully invest myself into exploring my father's suggestions. It wasn't until two years later that I met a black woman in her 40s who had had fibroid tumors. She had a PhD in organizational development, with emphasis in black feminism and anti-racism. We had the kind of spiritually gratifying conversations about racism and sexism that I rarely, if ever, had at Harvard, where I was attending part time, earning my Masters degree in Educational Technologies. She told me, "You need to get a book called *Sacred Woman* by Queen Afua. She can help you with your womb health."

Several days later, I ordered Queen Afua's *Sacred Woman* book. After reading the first half of the book, I was given a glimpse into the world of a vegan praxis that was rooted in race and gender consciousness with the collectivity of black females in mind. Queen Afua wrote from a black female racialized consciousness that spoke assertively to my own lived realities. She writes about how trauma and racialized colonialism affect the womb health of Black women, and how a well-planned vegan or plant-based diet can remedy this:

> I cry a river of tears that heal for the Negro slave woman, my great-great-grandmother, who was forced to part her thighs for the entrance of a pale pink penis to fulfill her owner's demonic quest to force his way violently into her soft dark womb, leaving his ... pardon me, I can't breathe, I'm still enraged two hundred years later. I still hurt. I still bleed. I'm outraged, feeling fear and helplessness for all my great-great-grandmothers who passed their self-hate, lack of self-esteem, their acceptance of abuse, their internal war down through the bloodline to me ... (Afua 2000, 57-58)

The quote above is integral to Queen Afua's over thirty years of health and food activism in the black female community, which is dedicated to teaching woman how the physical and psychic pain of the legacies of slavery are embedded in the memory of our minds and cells; that we must be conscious of not eating "standard American diet" comfort foods to deal with such racialized-sexualized pain and trauma, because these foods end up killing our womb health, breast health, spiritual health, total bodily health. Her message of veganism was rooted in de-colonial body politics and uplifting the black community to empower themselves against neo-colonialism and racism, first and foremost; a *race conscious* logos I *never* heard from "race-neutral" vegans, vegetarians, and animal rights activists at Dartmouth College.

What truly moved me into practicing veganism was not only reading Queen Afua, but also the words of African American comedian, activist, and vegan Dick Gregory. Dick Gregory, notes, quote:

> I personally would say that the quickest way to wipe out a group of people is
> to put them on a soul food diet. One of the tragedies is that the very folks in the
> black community who are most sophisticated in terms of the political realities
> in this country are nonetheless advocates of 'soul food.' They will lay down a
> heavy rap on genocide in America with regard to black folks, then walk into a
> soul food restaurant and help the genocide along. (Witt 2004, 133-134)

It was with the help of these two critical thinkers that I finally saw my own womb
ill-health as a symptom of systemic racism, sexism, nonhuman animal exploitation,
and corporate capitalism that became interconnected with my consumption of
nutritionally vapid and highly processed foods. Immediately, I made the transition
into a whole foods plant-based regiment. I specifically followed Queen Afua's
nutritional wisdom in *Sacred Woman*, and shrank my fibroid tumors by 75%. I
also followed her advice to eliminate not just "junk" food from my diet, but junk
energies that caused me great suffering; for her, consumption went beyond what
we ate; she advocated that her readers not consume negative energies such as
gossip, bad relationships, and television.

The Sistah Vegan project and book were born out of my desire to find out
how and why black females enter whole foods veganism from *our own lived
experiences* with what it has meant to be black and female in the USA. I know
there are black women who practice veganism for animal rights first, however, the
Sistah Vegan project over the last six years has revealed that a significant number
of black women ultimately chose a vegan lifestyle as a decolonial and anti-racist
method to combat health disparities in the black community; veganism is not the
only way, but it is one way that a group of black women have chosen to understand
anti-racist health activism. From that entry point, many of us eventually saw why
and how exploitation of non-human animal and their suffering and pain was *our*
suffering and pain. Some of us realized that the environmental racism occurring
in our communities, such as the fact that large meat production factories produce
pollution that are disproportionately found in poor communities of color because,
"Well, hey, they're just BLACK POOR people. Who gives a damn about them."
Therefore, not only did eating an over abundance of hormone injected, antibiotic
injected animal products for comfort food cause us black women to have higher
rates of fibroid tumors than the norm (Brown and Ifeyani 2002), or cause us to
have rising rates of colon cancer and diabetes in the black communities, we were
buying these products from corporations that support the dumping of animal
byproduct toxins into our own communities; corporations who target us with hip
hop oriented Burger King and KFC adds (Freeman 2007), *despite* knowing full
well that consumption of these products will never help resolve the black health
disparities or the fact that many black folk like myself, overdosed on Chicken
McNuggets to deal with racism. And many of us eventually realized that such
corporations were exploiting pigs, chickens, and cows in a frighteningly similar
way to African slave women in antebellum USA; an era when African women's
reproductive gifts were literally raped to produced human slaves, many of which

were taken away from these African mothers and never seen again. Many of these Sistah Vegan contributors saw the same agony from the mother cow who's baby is taken away to make 'veal,' or the mother cow who is constantly kept pregnant and babies taken away so she can produce milk for the human population. There was a time in the USA when we were raped, our babies stolen and sold away, and we were forced to breastfeed the slavemaster's children. Contributors in Sistah Vegan note that for them, this is one of the reasons that the animal rights message finally connected to their souls; however, for others, eating a plant-based diet had nothing to do with animal rights, simply the right to better human health and to avoid nutritional related ailments (Harper 2010).

What the Sistah Vegan project has done for me, is to help me better articulate how post traumatic stress around unresolved issues, and a lack of healing of racisms and other legacies of colonialism on the black female body, affect our health and psyche, and also why certain vegan outreach methods work better for certain groups over others. The Sistah Vegan project explores how racialized stress becomes both a challenge and an unexpected blessing for black female vegans. It is a challenge when many of us interact with mainstream vegan activists who have never viscerally experienced the consequences of racism or have never been made aware of how whiteness has racialized their vegan consciousness. However, I also feel strongly that black female vegans in the USA are able to meet these challenges because we are able to offer a culturally specific mode of veganism that reflects our own lived experiences of racism-sexism. Queen Afua (Afua 2000; Afua 2008;), Tracye McQuirter (McQuirter 2010) and Afya Ibomu (Ibomu 2008) are excellent examples of African American female vegans doing just that. Sistah Vegan clearly demonstrates that despite racisms, we are not helpless victims, we are survivors.

These days, I now know how to better handle the trauma and stress of racism, at least at the nutritional level. When I recently learned about the murder of Trayvon Martin, I was overwhelmed with anxiety and fear about the future of my own twin brother and my three-year old son. And similar to my first day of seventh grade when that boy called me "nigger," I wanted to reach for a nutritionally vapid comfort food. However, having had exposure to critical thinkers such as bell hooks, Queen Afua, and Dick Gregory, I reminded myself that overdosing on a bag of jelly-beans was not the answer (granted, they were organic and vegan, but still, 'sugar is still sugar'). The answer to my feeling hopelessness would lie in being compassionate towards my own anguish and my body's needs to survive through the psychic stress of hearing about the tragedy of Trayvon Martin (and so many like him). I put my vegan jelly-beans away, and opened the refrigerator to find a few leaves of kale that would be part of a green smoothie I would make in my blender. Potentially needing to stay in bed all day, sick because I had put my body out of harmony by eating 94 jelly-beans would not help anybody. If I was going to remain strong enough to continue doing the decolonizing work of the Sistah Vegan Project, then I needed to avoid any nutritionally induced ailments that could prohibit me from helping to prevent racialized violence. Queen Afua and Bryant Terry would be proud.

## References

Adams, Carol J. *Neither Man nor Beast: Feminism and the Defense of Animals.* New York: Continuum, 1994.

Afua, Queen. *Sacred Woman: A Guide to Healing the Feminine Body, Mind, and Spirit.* New York: Ballantine Publishing Group, 2000.

Afua, Queen. *The City of Wellness: Restoring Your Health through the Seven Kitchens of Consciousness.* Brooklyn: Queen Afua Wellness Institute Press, 2008.

Barnard, Dr. Neal. *Food for Life: How the New Four Food Groups Can Save Your Life.* New York: Random House, 1993.

Brown, Monique, and Ifeanyi C.O. Obiakor M.D. *It's a Sistah Thing: A Guide to Understanding and Dealing with Fibroids for Black Women*: Kensington, 2002.

Bullard, Robert D. *Confronting Environmental Racism: Voices from the Grassroots.* 1st ed. Boston, Mass.: South End Press, 1993.

Cvetkovich, Ann, and ebrary Inc. *An Archive of Feelings Trauma, Sexuality, and Lesbian Public Cultures,* Series Q. Durham, NC: Duke University Press, 2003.

Freeman, Andrea. "Fast Food: Oppression through Poor Nutrition." *California Law Review* 95, no. 2221 (2007): 2221-59.

Gottlieb, Robert, and Anupama Joshi. *Food Justice.* Cambridge: MIT Press, 2010.

Harper, A. Breeze. *Sistah Vegan: Black Female Vegans Speak on Food, Identity, Health, and Society.* New York: Lantern Books, 2010.

Harper, Amie Breeze. "Going Beyond the Normative, White, "Post-Racial" Vegan Epistemology." In *Taking Food Public: Redefining Foodways in a Changing World,* edited by Psyche Williams-Forson and Carole Counihan. New York: Routledge, 2011.

Ibomu, Afya. *The Vegan Soulfood Guide to the Galaxy.* Atlanta, Georgia: Nattral Unlimited, 2008.

Liu, Yvonne Yen, and Dr. Dominique Apollon. "The Color of Food." 26. Oakland, CA: Applied Research Center, February 2011.

McQuirter, Tracye Lynn. *By Any Greens Necessary: A Revolutionary Guide for Black Women Who Want to Eat Great, Get Healthy, Lose Weight, and Look Phat.* Chicago: Lawrence Hill Books, 2010.

Oliver, Kelly. *The Colonization of Psychic Space: A Psychoanalytic Social Theory of Oppression.* Minneapolis: University of Minnesota Press, 2004.

Randall, Vernellia. *Dying While Black: An Indepth Look at a Crisis in the American Healthcare System.* 1st ed. Dayton: Seven Principles Press, Inc, 2006.

Terry, Bryant. *Vegan Soul Kitchen: Fresh, Healthy, and Creative African-American Cuisine.* 1st ed. Cambridge: De Capo Press, 2009.

Torres, Bob. *Making a Killing: The Political Economy of Animal Rights*. Oakland, CA ; Edinburgh, Scotland: AK Press, 2008.

Williams-Forson, Psyche. *Building Houses out of Chicken Legs: Black Women, Food, & Power*. Chapel Hill: The University of North Carolina Press, 2006.

Witt, Doris. *Black Hunger: Soul Food and America*. Minneapolis: University of Minnesota Press, 2004.

# Thematic Tabs for Chapter 8

Colonial

Discourse

Race

Science

Chapter 8

# Traditional Knowledge and the Other in Alternative Dietary Advice

Edmund M. Harris

**Editors' Note**: This chapter engages with the themes of *colonial*, *discourse*, *race*, and *science*. Harris examines Weston A. Price Foundation (WAPF), and specifically Sally Fallon's "Ancient Dietary Wisdom for Tomorrow's Children" (included in this volume), which argues for the importance of traditional and indigenous diets and serves as a challenge to hegemonic nutrition. Harris notes that the WAPF unintentionally reproduces problematic and oppressive racial and colonial discourses in arguing for the importance of traditional or indigenous dietary knowledge. In exploring how the WAPF succeeds and falters in its resistance to hegemonic nutrition, Harris opens a discussion about how best to valorize traditional knowledge in a society that places great worth on ways of knowing based in Western science.

## Introduction

In the contentious world of nutrition advice and dietary interventions, few mount as direct a challenge to the nutritional mainstream as Sally Fallon Morrell and the Weston A. Price Foundation. Fallon's opposition to "politically correct nutrition" and advocacy of a diet based on "traditional wisdom" is based on research carried out during the first half of the twentieth century by Cleveland dentist Weston A. Price (1870–1948), as described in *Ancient Dietary Wisdom for Tomorrow's Children* (Fallon Morell 1999, reprinted in this volume). In this commentary, I focus on the ways in which *Ancient Dietary Wisdom for Tomorrow's Children* succeeds, and falters, in challenging the hegemonic discourse surrounding dietary advice for Fallon's primary audience in the United States. The traditional dietary model advocated by Fallon is appealing to many, but rests on a problematic foundation of dichotomous western thought that establishes divisions between nature and culture, primitive and modern, tradition and science. Price's work, as was common at the time, represents the primitive as morally virtuous and living close to nature, and sees primitive diets not only as a route toward healthier nutrition, but as a way to avoid the moral decline of western civilization. Fallon's writing remains remarkably close to this colonial vision. I argue that a strong challenge to hegemonic nutritional discourse needs not only to engage eaters in thinking beyond the confines of hegemonic nutrition (as Fallon does so successfully), but also to avoid reproducing oppressive discourses of race, gender and coloniality. I also compare Fallon's dietary politics to those of another advocate of a return to traditional diets: Michael Pollan. Their differing ways of knowing traditional diets make clear two distinct approaches to formulating a contemporary understanding of traditional dietary practice, one that expresses traditional knowledge in the language of western nutritional science, and another that is open to knowing traditional knowledge in a different register. The possibility of *knowing* nutrition differently remains another significant challenge in alternative dietary politics.

*Ancient Dietary Wisdom for Tomorrow's Children* (hereafter *ADWTC*) lays out an approach to diet based on the work of Weston A. Price that informs the goals of The Weston A. Price Foundation (WAPF), of which Fallon is President. Price's original research, presented in *Nutrition and Physical Degeneration* (1939), offers detailed descriptions of the diets of "isolated remnants of primitive racial stocks" (1939: 1) around the world, selected as the control populations in a research agenda that sought to demonstrate the damage to human health caused by the low nutrient density of foods in the westernized diet of the United States. Price argued that this western diet, which contained increasing proportions of industrially processed foods and sugar, represented the latest stage in a nutrition transition (Cabellero and Popkin 2002), the beginning of which could still be examined in isolated communities that had minimal contact with western civilization and its accompanying western diet that accompanied colonization.

Fallon's advocacy of a traditional diet represents a marginal, yet increasingly influential voice in the crowded arena of contemporary dietary advice. The work of agrifood scholars and the prevalence of popular dietary advice across all sectors of the print and electronic media paint a picture of consumers who are anxious about eating and are looking for dietary guidance (Jackson 2010, Levenstein 2003). In addition to official government-issued dietary guidelines (USDA CNPP 2010), US consumers are also surrounded by other contradictory influences, including easily accessible industrially produced, highly processed food, high carbohydrate diets, high protein diets, functional foods, diets that cut certain nutrients (e.g. gluten), vegetarianism, veganism, and advice from friends, family, and medical professionals. Uncertainty about what to eat is motivated by concerns about health, body image, and the wider impacts of food purchasing decisions (in environmental and socioeconomic terms), and is embedded in complex layers of social meaning relating to the food items themselves, the identities of consumers, and visceral fears about the ways in which food items might change the consumer's body upon consumption.

These meanings are captured in popular discourse around "eating well" or "eating right." While official dietary guidelines assume that individuals take a rational approach to food consumption based on their understandings of health outcomes, it is clear that eating is heavily influenced by social discourses of gendered bodily aesthetics and beauty, and the perceived "naturalness" of foods (leading to classifications of foods as "real" or "fake"). As such, individuals are interpellated through their eating decisions into subjectivities within a moral discourse of food consumption. In this discursive arena, individuals who do not follow the strictures of hegemonic nutritional guidance—or "good nutrition" (Crotty 1995)—are subject to blame and guilt. The combination of scientific dietary advice promulgated by experts and the moralizing discourse of individual culpability and fault elide the structural factors that impact the availability of different foods, and the acceptability of different approaches to "right nutrition."

*Ancient Dietary Wisdom for Tomorrow's Children* argues for the contemporary relevance of Price's research as the foundation for a different approach to diet and nutrition that challenges the mainstream, or hegemonic, nutritional discourse in the United States. Price hypothesized that the "facial deformities" caused by poor dental health that he saw among his younger clients were the product of changes in diet, and embarked on an international program of field research in order to study the diets of populations who ate a non-western diet. Such populations, described by Price as "primitive racial groups" (1939: xix) provided the experimental control group, whose diet could be compared to those in the United States. Price counted the dental caries present in each group he visited, recorded the characteristics of their diet, and took photographs of his research subjects' faces, often with wide smiles to show the quality of their teeth. Fallon describes how Price "found tribes or villages where virtually every individual exhibited genuine physical perfection" and noted that "the natives were invariably cheerful and optimistic" (Fallon Morell 1999).

Price rejected the argument that the good health of his research subjects was due to "racial purity" (a popular opinion at the time), arguing instead that incidences of poor health were determined by contact with traders or missionaries (or other representatives of western colonialism) and the resulting prevalence of the "displacing foods of modern commerce" (Fallon Morell 1999). The diets of the groups described in Price's research vary dramatically, but Price argues that they share certain features that Fallon and the Weston A. Price Foundation mobilize as the basis for their alternative dietary message:

> The foods that allow people of every race and climate to be healthy are whole natural foods – meat with its fat, organ meats, whole milk products, fish, insects, whole grains, tubers, vegetables and fruit – not newfangled concoctions made with white sugar, refined flour and rancid and chemically altered vegetable oils. (Fallon Morell 1999)

Subsequent laboratory analysis carried out by Price demonstrated that such "whole, natural foods" contained significantly higher levels of minerals and fat-soluble vitamins, particularly vitamins A and D which are primarily found in animal fats, organ meats and fish oils. Fallon also recounts Price's discovery of a new vitamin-like activating substance that affects nutrient absorption and the healing of dental caries and fractured bones (Price 1939: 404-5). Price termed this substance "Activator X" and demonstrated its singular presence in pasture-raised meat, leading the Weston A. Price Foundation to advocate the consumption of pasture-raised meat and dairy products. For Fallon, the findings of Price's research are indisputable, although they have been largely ignored by hegemonic, mainstream nutritional science, which maintains that consumption of saturated fats and cholesterol from animal sources should be minimized or avoided altogether (USDA CNPP 2010).

Fallon then continues to review a series of trends in nutrition research that support Price's findings, including the increased attention to the roles played by vitamins, fish oils, saturated animal fats and cholesterol in the healthy functioning of a variety of biophysical processes. This growing body of evidence challenges dominant nutritional science which focuses on the link between saturated fat, cholesterol and heart disease (the "lipid hypothesis"), first proposed in the 1950s and still central to mainstream dietary advice today (Willett 2001: 58, Taubes 2007). These recent trends in nutritional research indicate that replacing fat and cholesterol with carbohydrates and vegetable oils has caused widespread negative health consequences. Fallon concludes by re-articulating Price's enthusiasm for modeling US diets on those of primitive groups, arguing that "the wisdom of the so-called primitive" has much to teach us both about diet and sociocultural practices that lead to good health.

This alternative approach to diet and nutrition now informs the advocacy of the Weston A. Price Foundation (http://www.westonaprice.org/), which provides individuals with the resources to follow a "traditional diet" and conducts advocacy around policy issues including the availability of raw milk (http://www.realmilk.com/), nutrition legislation, the risks of soy products, and other issues. Foundation membership is organized in local chapters with a total of 10,500 members in 2008, when the Foundation had gross revenues of $1.213 million (Black 2008, WAPF 2011).

**Diet, Morals, and the Primitive Other**

Fallon and the Weston A. Price Foundation argue for a diet based on the traditional knowledge and practices of indigenous people of the Global South. This approach focuses on the benefits of traditional diets and relies heavily on representations of "the primitive" that are deeply problematic in the context of postcolonial critique (Said 1978). Fallon is not alone, however, in drawing on an idealized construction of the primitive as a way to think about alternatives to modern, Western lifestyles. These essentialized imaginaries of the primitive emerged through the Western knowledge systems of the colonial era, representing both a people to be feared and controlled, and a gentle people living in harmony with nature.

Although Fallon occasionally uses scare quotes[1] around "primitive" to indicate an awareness of the problematic nature of the term, much of her (and other

1   Following Torgovnick (1999, 20), I have not put *primitive* into quotation marks in my writing. While such marks are valuable in reminding the reader of the constructed nature of the term and raise the idea that the primitive does not or has never existed in an

Weston A. Price Foundation members') writing reproduces the construction of the primitive as a noble savage living in harmony with nature. However the trope of the noble savage represents just one side of a binary that has emerged through scientific, literary and artistic interpretations of encounters with indigenous people during European colonial expansion:

> [The primitives] exist for us in a cherished series of dichotomies: by turns gentle, in tune with nature, paradisal, ideal – or violent, in need of control; what we should emulate or, alternatively, what we should fear; noble savages or cannibals. (Torgovnick 1990: 3)

While for Price the primitive was to be emulated, both images have played important roles in the construction of Western discourses of the self. Said has demonstrated how the construction of identity "involves establishing opposites and 'others' whose actuality is always subject to the continuous interpretation and re-interpretation of their differences from 'us'" (1978: 322). In the case of white Americans or Europeans, these "others" are often variants either of the violent or the noble savage, to be controlled or emulated respectively, but never to be fully embraced. The presence of alterity, whether actual or constructed, seems necessary to maintain our efforts to understand and assert western identities (Fabian 2006: 148), and yet this relationship is paradoxical:

> Modern thought is pledged to a kind of applied Hegelianism: seeking its Self in its Other ... Modern sensibility moves between two seemingly contradictory but actually related impulses: surrender to the exotic, the strange, the other; and the domestication of the exotic, chiefly through science. (Sontag 1964: 69-70)

The discursive construction of the primitive other that took place as western colonizers sought to understand indigenous people around the world has a strong temporal component, in which the primitive is positioned as existing in an earlier stage of development than Europeans or Americans. Initially, this temporal difference was expressed in evolutionary terms with non-white groups considered closer to our non-human evolutionary ancestors. In the latter half of the twentieth century, this perspective shifted toward a temporality based on economic development and the degree to which indigenous groups have been incorporated into the global capitalist economy. In both cases, however, representations of the primitive produce discourses of *temporal* and *developmental* difference to explain what were, and still are, *spatial* and *cultural* differences. Doreen Massey, a geographer who has theorized the production of space, argues that this discursive

---

objective form, there are then any number of other terms in this text that also deserve to be flagged as constructs in a similar way, such *west* or *tradition*. Rather, I have used quotation marks only when citing an idea or phrase in another text, and rely upon the content to make clear that *primitive* is an ideological construct with a variety of cultural meanings.

sleight of hand turns geography into history and has egregious social and political effects, creating a narrative in which the others (nations or people) of the Global South "are not recognized as coeval others ... [but are] merely at an earlier stage in the one and only narrative that it is possible to tell" (2005: 5).

The appeal to the diet of a primitive other echoes feminist theorist bell hooks' analysis of the way in which constructions of the other are approached and incorporated into western subjectivities through consumption, or by "eating the Other" (hooks 1992: 21). Through consumption, hooks argues, the difference between the westerner and the primitive is reproduced in commodities such as food, drawing on a "contemporary longing for the 'primitive' [that] is expressed by the projection onto the Other of a sense of plenty, bounty, a field of dreams" (hooks 1992: 25). Western consumers partake in this consumption of the Other, hooks suggests, while simultaneously seeking to maintain their sense of self as distinct from the Other: "one desires contact with the Other even as one wishes boundaries to remain intact" (1992: 29). This paradoxical engagement with the other through food consumption is most clear in the enthusiasm for foreign or "exotic" cuisine in contemporary Anglo-American culture, in pursuit of a multicultural, adventurous, or cosmopolitan lifestyle (Heldke 2003). Fallon's traditional diet also represents a desire to consume the other, although in pursuit of a different set of values organized around moral virtue and proximity to what is "natural." The contradictory needs to incorporate the other while also maintaining its difference, as described by hooks, are also expressed by Fallon's desire to embrace primitive or traditional knowledge while simultaneously expressing this knowledge in western scientific terms.

As a 1930s forerunner of medical anthropology, it is not surprising that Price interpreted his research subjects as "primitives," since this trope was widespread in western thought about indigenous people in the Global South at the time. In this sense, however well intentioned, Price's research contributed to the singular narrative of the non-western other as fundamentally different and living in an earlier stage of development, closer to nature. Although Price saw a learning opportunity in his contact with "primitives" for western populations who had strayed too far from nature's (and also God's) intended path, his research remains part of the Euro-American discourse of indigenous otherness that perpetuated the injustices of the colonial system. In seems unrealistic, therefore, to argue that Price's choice of research subjects was solely concerned with indigenous groups' dietary practices, and was not also guided by the imaginaries of the primitive that were prevalent at the time, and in particular, the image of the noble savage living close to nature. In response to the (rhetorical) question "Why seek wisdom from primitive people" posed at the start of his book, Price states:

> [I am] fully aware that [my] message is not orthodox; but since our orthodox theories have not saved us we may have to readjust them to bring them into harmony with nature's laws. Nature must be obeyed, not orthodoxy. Apparently many primitive races have understood her language better than have our

> modernized groups … A decadent [western] individual cannot regenerate himself, although he can reduce the progressive decadence in the next generation, or can vastly improve that generation, by using the demonstrated wisdom of the primitive races. (1939: 6-7)

This response exemplifies the prominent nature/culture dichotomy that underlies Price's work, and excerpts like this demonstrate how deeply Price's research was embedded the colonial ontology of nature, culture, and development. While Price's attitude should be read in this historical context, it is surprising that Fallon and WAPF do not seek to distance themselves more from the discourse of otherness based on the imaginary of the primitive living close to nature. On reading *Ancient Dietary Wisdom for Tomorrow's Children*, it seems that the Weston A. Price Foundation still *needs* "the primitive" as imagined by Price to articulate its alternative dietary politics. The relative proximity of Fallon's message to this oppressive colonial discourse seems like a missed opportunity to acknowledge the uneven power relations of research conducted *on* (rather than *with*) indigenous groups and the continued salience of constructions of nature in guiding consumption choices.

The passage quoted above also illustrates how Price saw a reconnection with nature as the antidote to the decline and decadence of modern, industrialized western society. This argument reveals the normative tone of Price's work, and the reference to decadence rather than simply to ill-health demonstrates the implied relationship between diet and moral virtues: that by eating differently we might become better people. The links between food and morality reach back to antiquity, and center on the fact that the pleasure taken from food consumption challenges ideals of bodily self-control in a way that is paralleled only by sex (Coveney 2000: vii, Probyn 2000). In western society, where bodily appetites and desires were (and still are) to be controlled and suppressed (Foucault 1990), food consumption can generate significant anxiety, guilt and shame, and it is also the basis for a variety of moral judgments made about others. John Coveney, a professor of public health, addresses the link between food and morality, arguing that:

> scientific and technical knowledge forms the basis for the moral judgments we make about ourselves and others. It is this moral imperative which is encoded in nutrition that makes it so compelling, so engaging, so judgmental, and so strangely popular. (2000: viii)

Thus, Price's reference to decadence represents the nutritional component of a broader normative discourse that associates modern western lifestyles with a decline in moral values and a compromised approach to the control and regulation of the body and personal health. The rhetorical link between food and bodily control is also central to Fallon's alternative dietary politics, as demonstrated by a speech in response to the release of the 2010 dietary guidelines by the US Department of Agriculture (USDA):

> [The USDA] preaches a kind of low-fat, high-fiber, low-salt puritanical diet that is impossible to eat. We have cravings for these kinds of foods because we need these foods; we need saturated fats; we need salt in our diet ... and the result is that people end up eating what I call the pornographic foods [the presentation slide shows a range of highly processed snack foods]. (Fallon Morell 2011)

The USDA guidelines are framed here as a call for people to abstain from a consumption practice that is governed by bodily desire. When bodily desire then wins over puritanical rationality, people capitulate to a quick and easy form of satiation: "pornographic foods." The labeling of processed snack foods as pornographic draws an immediate equivalence between food and sex as activities that satisfy visceral desires and establish a distinction between the "wrong" way to satisfy these desires – processed food (and by implication, pornography), and the "right" way – whole foods (and sexual intercourse?). This food/sex equivalence rests on a discourse that positions bodily desires as properly subject to regulation by social norms, and suggests that the practice of consuming processed food should be conceptualized as equivalent to watching pornography: something one is ashamed of, and that never takes place in public.

This sequence of representations – puritanical dictate, followed by forbidden bodily desire, followed by capitulation to the pornographic – presents a paradoxical moralizing message. At first, the puritanical diet to be resisted: this is the approach of hegemonic nutritional science that denies consumers the nutrients their bodies need and desire. Fallon's counterargument against the puritanical dictates of hegemonic nutritional science rests here on the idea of *what our bodies need*, based on Price's argument that the body's *need* for animal fats is *natural*. Following this argument, however, Fallon then uses the term "pornographic" to describe the highly processed foods consumers eat to fulfill their bodily desires. This contradicts the rejection of Puritanism in the preceding sentence, since by labeling processed foods pornographic, Fallon draws a metaphorical equivalence between viewing (consuming) sexual images and eating processed foods, two bodily desires that Puritanism would deny. These competing references suggest that Fallon is not in favor of accepting *all* bodily desires – decadent western temptations such as pornography and processed foods are to be resisted, whereas other bodily desires, such as the need for saturated animal fats, are deemed natural and wholesome and should be embraced.

Price and Fallon's moral message also extends beyond the rights and wrongs of individual food consumption to address broader social problems. Price attributed a wide variety of social ills to poor diet, including delinquency, criminality and backwardness (1939: 17-20):

> The origin of personality and character appear in the light of the newer data to be biologic products and to a much less degree than usually considered pure hereditary traits. ... Mass behavior therefore ... becomes the result of natural forces, the expression of which may not be modified by propaganda but will

require correction at the source. Nature has been at this process of building human cultures through many millenniums and our culture has not only its own experience to draw from but that of parallel races living today as those that lived in the past. [Price's research], accordingly, includes data that have been obtained from several of Nature's other biologic experiments to throw light on the problems of our modern white civilization. (Price 1939: 4)

Writing 60 years later, this hypothesis of a causal link between good diet and the production of good citizens is echoed by Fallon:

For a future of healthy children—for any future at all—we must turn our backs on the dietary advice of sophisticated medical orthodoxy and return to the food wisdom of our so-called primitive ancestors, choosing traditional whole foods ... When offspring are properly spaced, and care given to the diet of both parents before conception, and to the children during their period of growth and development, all children in the family can be blessed with the kind of good health that allows them a carefree childhood; and the energy and intelligence they need to put their adult years to best and highest use. (Fallon Morell 1999)

In these closing paragraphs, Fallon reiterates the need to reject the dictates of hegemonic nutrition, and return to a traditional diet. This is not just advice about what to eat, but normative advice about lifestyle in the face of the corrupting forces of modern industrial society. The moral tones are clear in the connection of nutritional choices to correct modes of parenting, normative family structures, and the ability of children to becoming good and productive citizens.

### Valuing Traditional Knowledges: A Comparison with Michael Pollan

The appeal to "traditional diets" is the cornerstone of Fallon's alternative nutritional politics. This emphasis is shared by the popular food journalist Michael Pollan, who has authored numerous articles and books exploring alternative food politics. Fallon and Pollan's approaches to alternative dietary guidance both mobilize the imaginary of the healthy, traditional diet, and both reject mainstream hegemonic nutritional science. This comparison is revealing because, despite these similarities, they deploy different rhetoric regarding the "other" whose traditional diet we should emulate, and about the way that traditional knowledge should interact with hegemonic nutritional science. These differences are accentuated by Fallon's vigorous rejection of Pollan's work; differences that I argue arise from divergent ontological commitments to western scientific knowledge.

*In Defense of Food* (2008) presents Pollan's critique of hegemonic nutritional science, and an argument that dietary choices have been complicated by an approach to nutritional science that breaks foods down into their component nutrients rather than taking a holistic approach to diet. Pollan paints a picture of nutritional science as a relatively young science (2009: xi), which commits the reductionist error of ignoring dietary context, drawing on both Gyorgy Scrinis's notion of nutritionism (Scrinis 2008, see also Scrinis this volume), and Marion Nestle's warnings about the decontextualized nature of nutrition science:

> The problem with nutrient-by-nutrient nutrition science is that it takes the nutrient out of the context of the food, the food out of the context of the diet, and the diet out of the context of the lifestyle. (Nestle, quoted by Pollan 2008: 62)

Pollan reviews the often contradictory messages that result from nutritional studies, which have in turn singled out saturated fat, cholesterol and carbohydrates as causes of the diseases of Western civilization, and argues that each time one nutrient is declared to be "bad," diets shift to include more of other nutrients, which in turn cause different problems. This pattern is demonstrated by the nationwide response to the lipid hypothesis: as consumers cut foods high in fat (particularly red meat and dairy products) out of their diets, the caloric value of these foods was replaced by carbohydrates and "low-fat" products which are often highly processed or refined. Pollan argues that while this transition produced a reduction in saturated fat consumption, overall fat consumption held steady and caloric intake increased, an irony he blames on what Scrinis coined as the *ideology of nutritionism* (2008: 51):

> By framing dietary advice in terms of good and bad nutrients, and by burying the recommendation that we should eat less of any particular actual food, it was easy for the take-home message of the 1977 and 1982 dietary guidelines to be simplified as follows: *Eat more low-fat foods*. (Pollan 2008: 51, original emphasis)

Pollan rejects the nutritional scientific focus on including and excluding specific nutrients, which he argues is too often led by the demands of the food and medical industries, and looks instead to traditional diets for a holistic alternative to the modern western diet:

> I'm inclined to think that any traditional diet will do; if it wasn't a healthy regimen, the diet and the people that followed it wouldn't still be around. (Pollan 2008: 173)

While Pollan's message might seem very similar to that of the Weston A. Price Foundation, his books have all received negative reviews from Fallon and other WAPF website contributors. These reviews – of *Food Rules, The Omnivore's Dilemma* and *In Defense of Food* (Fallon Morrell 2010a, 2010b, Ussery 2009) – all criticize Pollan for advocating a plant-based diet, and for ignoring (or misrepresenting) Price's research that demonstrates the importance of consuming pasture-raised animal products. Fallon states that Pollan "comes out firmly against meat" (2010a) and Ussery refers to "Pollan's frequent refrain that we should 'eat less meat'" (2009), a phrase that in fact only appears in Pollan's book as a paraphrase of the government dietary guidelines that emerged from the McGovern Committee (2008: 51). By portraying Pollan as arguing against meat consumption, the Weston A. Price Foundation misrepresents his work: he does in fact advocate eating meat in moderation, and advises readers to focus on the conditions under which their meat is raised. Rule #27 in Pollan's *Food Rules* advises readers to "Eat animals that have themselves eaten well" (2009: 61) is an argument for pasture raised meat that is actually a good fit with the Weston A. Price Foundation position. What can explain the Weston A. Price Foundation's vigorous criticism of Pollan's work, when it seems that these proponents of alternative dietary practices have so much in common?

I suggest that the root of the discomfort felt by Fallon and her colleagues with Pollan's approach lies in his treatment of mainstream nutritional science, and in particular, his rejection of dietary advice (hegemonic or alternative) *that focuses on individual nutrients*. Price's research was a forerunner of modern nutritional research that breaks food down into its constituent components, and Fallon's primary political message concerns the rescue of the vilified nutrient *saturated fat* from the clutches of the lipid hypothesis.

In response to this criticism, Pollan has commented on the strength of opinions held around dietary advice:

> The Weston A. Price Foundation … are fierce in their love of animal fat. And with pastured animal fat, healthy animal fat, a lot of what they say is right. But they really don't like plants. People feel like they have to take sides on this plant/ animal divide, and I don't think we do … People have strong, quasi religious views on these things. Secularizing the issue is challenging. (Pollan, in interview with Jeffery 2009).

Pollan's response suggests that among alternative diet advocates, ontological starting points play a significant role in the expression of alternative dietary advice. The suggestion that Fallon's opposition to Pollan's more pluralist style is "quasi religious" does not seem unjustified given Fallon's degree of comfort with references to "God's plan for mankind" and "the creator" (*ADWTC*). Pollan, however, appears to use "religious" to stand in for intolerance of different viewpoints, and while Fallon certainly reacts strongly to Pollan's different approach, it seems that the underlying difference relates to the use and value of traditional and scientific knowledges, rather than religion. Fallon and Pollan each look to different "others" as the source of traditional dietary knowledge, and each approach the relationship between traditional knowledge and western science differently.

## Learning from Traditional Dietary Knowledge

The first point of difference in comparing these traditionalist approaches to dietary advice is the proposed source of traditional knowledge. For Fallon, as discussed above, the primitive populations surveyed by Price are the source of traditional dietary knowledge on which alternative diets should be based. Pollan, however, produces a traditional imaginary that is closer to home:

> Don't eat anything your great grandmother wouldn't recognize as food ... we need to go back at least a couple of generations, to a time before the advent of most modern foods. (Pollan 2008: 148).

Neither Fallon nor Pollan have first-hand knowledge of the traditional diets they advocate: both forms of alternative dietary advice rest on learning from an "other," either the primitive (distant in space and time), or our own ancestors (distant in time, if not in space). These "others" are also distinct in cultural terms. While Pollan's imagined "other" is of the eater's own culture, Fallon's advice rests on the imaginary of the culturally-different primitive. This cultural difference raises the question of the legibility of the traditional knowledge, since in Fallon's case it must be translated between cultures and socio-environmental contexts. However, Fallon's primitive "other" is compelling precisely because of its difference to, and distance from, Western civilization, with its accompanying diet, disease, and self-constructed sense of moral decline. By constructing the imagined "other" as one allied—through Western colonial discourse—with nature, learning from primitive diets also offers the possibility of reconnecting with nature.

Pollan's choice to imagine our great-grandmothers as a repository of traditional dietary knowledge, while avoiding the difficulties of cultural translation, raises its own problems. The choice to rely instead on a gendered historical narrative in which older women hold the dietary knowledge needed to reform our food habits reproduces a pernicious form of gendered politics, which is increasingly widespread in the alternative agrifood movement. In this narrative, the movement of many women into paid employment during the latter half of the twentieth century resulted in the loss of traditional food practices, which relied heavily on the unpaid domestic labor of women (Deutsch 2011). This gender analysis often remains implicit in Pollan's writing, which calls for all consumers to take a more active role in food shopping and preparation, but by default turns to women as a source of knowledge about "good" food:

> So whom did we rely on before the scientists ... began telling us how to eat? We relied of course on our mothers and grandmothers and more distant ancestors, which is another way of saying, on tradition and culture. (Pollan 2009: xvi)

This comparison demonstrates that both Fallon and Pollan's appeals to traditional diets rely on an imagined other in their alternative dietary advice.

In neither case is the other the actual source of dietary knowledge: the great grandmother simply serves as a representational vehicle for Pollan's alternative dietary advice, standing in for cultural heritage, in the same way that the close-to-nature primitive does for Fallon. It seems clear that such othering represents a significant failing in attempts to do nutrition outside the dominant discourses that reproduce oppression in western society. This critique, therefore, challenges alternative diet activists to present their advice in a way that does not reinforce or perpetuate existing inequalities founded on essentialized racial, gendered, colonial, or nature/culture categories.

The second comparison to draw between Fallon and Pollan concerns the way they approach the relationship between traditional knowledges and western scientific knowledge. Mainstream hegemonic dietary advice rests upon the epistemology and practices of nutritional science, produced by a community of "experts" who seek to discover the single, right way to eat for optimum health by balancing a range of individual nutrients. Both Fallon and Pollan reject the message of hegemonic nutrition science, which they argue is strongly shaped by the relationships between agricultural subsidies, commodity production, agribusiness and food processing corporations, and a food and agriculture regulatory system populated by industry insiders (Meghani and Kuzma 2011). On closer examination, however, Fallon and Pollan each adopt distinctive antagonistic stances toward hegemonic nutritional science.

Pollan argues that we can, to some degree, simply *ignore* "the crosscurrents of conflicting science" (2008: 140). Nutritional science is too compromised by the *ideology of nutritionism*—the conceptualization of foods as aggregations of individual nutrients—to provide balanced or holistic dietary advice that does not rest upon "food-like substances," heavily processed foods, artificially manipulated to improve the delivery of specific nutrients (Pollan 2008: 80, Scrinis 2008). Rather than following this reductionist scientific approach, Pollan advocates a withdrawal from nutritional science:

> scientists can argue all they want about the biological mechanisms behind this phenomenon, but whatever they are, the solution is simple: *Stop eating a Western diet.* (2008: 140-41)

The relationship between hegemonic nutritional science and the Weston A. Price Foundation is more complex. While Fallon argues for a return to traditional diets, she views nutritional science as key to demonstrating the value of such diets. Rather than rejecting nutritional science altogether, Fallon refers to "bad science" and "good science" (e.g. 2010b), and describes the Weston A. Price Foundation as "dedicated to providing accurate information about nutrition ... we pride ourselves on showing the *scientific validation of traditional foodways*" (Fallon Morell 2011, emphasis added). Fallon does not challenge the epistemological foundations of nutritional science, therefore, but seeks to generate "good science" to rescue the currently vilified nutrients—saturated fats in particular—that

Price's research suggests are essential to a healthy diet. While Pollan rejects the reductionist framework of nutritional science, Fallon seeks to work within it. As such, Fallon's rejection of Pollan's dietary politics may be a response to his analysis of the "ideology of nutritionism," since it could be argued that by seeking to demonstrate the value of saturated fats and other nutrients in a western scientific framework, WAPF is working within rather than outside the nutritionism ideological framework. Pollan has not made such an argument, however, and reserves his criticism of the ideology of nutritionism for mainstream nutritional science.

Within the landscape of alternative dietary politics, these distinct positions represent different strategies for challenging, and attempting to dismantle hegemonic dietary advice based on mainstream nutritional science. Fallon's political advocacy takes aim at government dietary guidelines, and at government regulations that either limit access to foods considered desirable by the Weston A. Price Foundation (e.g. raw milk), or that enable access to foods the Foundation considers dangerous (e.g. soy products). These attempts to shift the balance of power in agricultural regulation and dietary advice away from agribusiness and food processing corporations should be applauded, and represent a key success in the Weston A. Price Foundation's attempts to do nutrition differently. An unfortunate irony for Fallon is that despite her enthusiasm for achieving scientific validation for traditional diets, the WAPF has relatively little peer-reviewed scientific evidence that supports their claims. While the Foundation's broader arguments against processed foods and around alternatives to the lipid hypothesis can find some support among scientific studies (e.g. Stanley 2010, Astrup et al. 2011), Fallon and other Weston A. Price Foundation members continue to base many of their dietary claims upon Price's original research. Those criticized by the Weston A. Price Foundation have highlighted this shortage of peer-reviewed evidence and have questioned the scientific credentials of Fallon and other WAPF writers, suggesting that establishing the WAPF message within the discourse of nutritional science remains a challenge.[2]

The comparisons of Fallon and Pollan's advocacy for traditional diets presented here raise the question of how best to valorize traditional knowledge in a society that places high value on scientific knowledge. The distinction between traditional and scientific knowledge is often conceptualized according to the identity of the knowledge-holder: traditional or "lay" knowledge is acquired by an individual through experience and with no formal or professional training, whereas scientific or "expert" knowledge is held by an individual with professional training (Collins and Evans 2007). The epistemology of western science positions such expert

---

2  The exchange between Christopher Masterjohn (2005, 2007) (a Weston A. Price Foundation member) and T. Colin Campbell (2006) (author of *The China Study*) serves to illustrate the type of debate that is prevalent in web forums and discussion boards, in which WAPF members defend their views based on Price's work, and are criticized for a lack of scientific credentials and for their over-reliance on a 70 year old scientific study.

knowledge as universally applicable, whereas traditional or lay knowledge is often conceptualized as local and contextual because of its experiential nature.

Both Fallon and Pollan advocate looking beyond the boundaries of expert nutritional science for new approaches to a difficult problem – the unhealthy western diet. In their references to traditional dietary knowledge, they each describe a lay or non-scientific knowledge (or wisdom) of healthy food sources and practices developed over generations of experience, rather than through the reductionist scientific method of western nutritional science. While Pollan is content to reference this traditional knowledge as a cultural source, Fallon seeks to translate traditional, non-western knowledge into western scientific terms. Indeed, the Weston A. Price Foundation mission statement articulates this translation as a central goal:

> [We] stand united in the belief that modern technology should be harnessed as a servant to the wise and nurturing traditions of our ancestors rather than used as a force destructive to the environment and human health; and that science and knowledge can validate those traditions. (WAPF 2000)

The mission to translate, valorize, and prove the truth of traditional dietary knowledges using the western scientific method can be interpreted in different ways. These efforts could be read as part of the broader effort to embrace traditional knowledges, and to decolonize dietary practices that have previously been marginalized (Milburn 2004). Wilson argues that projects emphasizing indigenous knowledges can challenge "the powerful institutions of colonization that have routinely dismissed alternative knowledges and ways of being as irrelevant to the modern world" (2004: 359). However, Fallon's unwillingness to accept traditional dietary knowledge on its own terms—as another way of knowing—may signal the continued colonial hegemony of Western science in the appropriation of indigenous knowledges today.

It is in this question of *how to know* traditional diets where the tension lies between Fallon and Pollan. By seeking to validate traditional dietary knowledge through the practices and institutions of Western science, the Weston A. Price Foundation hopes to speak—in the future—to a much larger audience; an audience that wants dietary advice to be supported scientific evidence. This desire to fundamentally reorient nutritional science represents a powerful, if as yet unrealized, challenge to the "hegemonic nutrition" that guides the current mainstream approach to diet and eating. This approach, however, would reproduce eaters' current dependence on scientific experts for dietary advice, whereas Pollan argues for an approach to diet in which we no longer need scientists to tell us what to eat:

> I am skeptical of a lot of what passes for nutritional science, and I believe there are other sources of wisdom in the world and other vocabularies in which to talk intelligently about food. (2009: xvi)

It is Pollan's hope that we can rediscover a common-sense approach to eating based on shared cultural heritage and a lay knowledge of "good nutrition." This viewpoint represents a move toward an acceptance of different ways of knowing in dietary practice, even if it remains couched in a romanticized image of the past and a set of troubling gendered stereotypes.

### Future Challenges for Alternative Dietary Politics

This commentary demonstrates the successes and growing influence of the Weston A. Price Foundation in challenging the hegemonic approach to nutrition and diet in the United States. In particular, Fallon's leadership in opposing regulation that limits the sale of raw milk and pasture-raised meats by small-scale farmers has been important to many in recent years. Moreover, Fallon's activism has introduced more than 10,000 consumers to a different approach to dietary advice, and has highlighted the extent to which our dietary choices are structured by a specific set of dietary guidelines formulated in a way that fits closely the agendas of the corporate agribusiness sector.

I have also argued here, however, that this powerful activism is founded on a series of unexamined assumptions and oppressive racial tropes concerning "primitive diets" that have been adopted in an unreflexive fashion from Price's 1930s research. In the analysis above, I have sought to demonstrate how these imaginaries of the primitive are expressed in *Ancient Dietary Wisdom for Tomorrow's Children*, and to highlight the way they shape Fallon's alternative dietary politics. Through comparison with Pollan's work, I have made clear that other alternative dietary approaches also rest on assumptions about those that hold "good" dietary knowledge, and have also teased out some distinctions between Fallon's and Pollan's work concerning their divergent attitudes toward Western nutritional science. By pulling at some of the discursive threads that caught my attention as I read *ADWTC*, I have demonstrated the ways in which alternative dietary politics often remains entangled in the dichotomous systems of thought that have characterized the era of western coloniality: nature and culture, primitive and modern, tradition and science. I conclude by noting two specific challenges that alternative dietary politics must still overcome as we develop more equitable ways of knowing, talking about, and eating food.

The reliance on the imaginary of the primitive—healthy, virtuous, and close to nature—is the most problematic discursive thread in *ADWTC*. This import from Price's research brings with it a host of essentializing baggage concerning "natural" and "unnatural" diets, racial stereotypes, and discourses of evolutionary and economic development that would be best left in research reports from the 1930s. If we are to look to other cultures for guidance about how to reorient our dietary and nutritional practices, it is important that we recognize the violence with which "other" knowledges have been appropriated in the past, and work to ensure such learning processes are equitable and part of the decolonization of diets, rather than a retrenchment of colonial power structures.

The comparison with Pollan's work demonstrates the diversity within calls to rediscover traditional diets, and highlights a division between approaches that seek to challenge hegemonic nutrition from within and those that reject nutritional science and its underlying ideology altogether. These divergent dietary politics produce different ways of knowing traditional diets, either through translation into western scientific knowledge, or through an exploration of different registers that might allow us to talk about nutrition without compromising a holistic understanding of diet and eating. Whatever the approach to learning from traditional dietary knowledges, doing nutrition differently requires that we move away from the guilt and blame associated with singular explanations of the "right" way to eat and what *the* "healthy" body looks like, and become open to a greater diversity of ways to "eat right."

## References

Astrup, A., Dyerberg, J., Elwood, P., Hermansen, K., Hu, F.B., Jakobsen, M.U., Kok, F.J., Krauss, R.M., Lecerf, J.M., LeGrand, P., Nestel, P., Riserus, U., Sanders, T., Sinclair, A., Stender, S., Tholstrup, T. and Willett, W.C. 2011. The role of reducing intakes of saturated fat in the prevention of cardiovascular disease: where does the evidence stand in 2010? *American Journal of Clinical Nutrition* [Online]. Available at: http://www.ajcn.org/content/early/2011/01/26/ ajcn.110.004622.abstract [accessed: 1 March 2011].

Black, J. 2008. The great divide: who says good nutrition means animal fats? Weston A. Price. *The Washington Post* [Online, 6 August] Available at: http://www. washingtonpost.com/wp-dyn/content/article/2008/08/05/AR2008080501000. html [accessed: 1 March 2011].

Cabellero, B. and Popkin, B.M. (eds) 2002. *The Nutrition Transition: Diet and Disease in the Developing World*. London: Elsevier Science.

Campbell, T.C. 2006. T. Colin Campbell's response to questions raised about the book, "The China Study. Startling Implications for Diet, Weight Loss and Long-Term Health." *VegSource* [Online, 1 October] Available at: http://www. vegsource.com/articles2/campbell_china_response.htm [accessed: 1 March 2011].

Collins, H. and Evans, R. 2007. *Rethinking Expertise*. Chicago: University of Chicago Press.

Coveney, J. 2000. *Food, Morals and Meaning: The Pleasure and Anxiety of Eating*. London: Routledge.

Crotty, P. 1995. *Good Nutrition? Fact and Fashion in Dietary Advice*. Sydney: Allen and Unwin.

Deutsch, T. 2011. Memories of mothers in the kitchen: local foods, history, and women's work. *Radical History Review*, 110, 167–177.

Fabian, J. 2006. The other revisited: critical afterthoughts. *Anthropological Theory*, 6(2), 139–52.

Fallon Morell, S. 1999. *Ancient dietary wisdom for tomorrow's children* [Online: Weston A. Price Foundation]. Available at: http://www.westonaprice.org/traditional-diets/622-ancient-dietary-wisdom.html [accessed: 1 March 2011].

Fallon Morell, S. 2010a. *Food Rules by Michael Pollan.* [Online: Weston A. Price Foundation]. Available at: https://westonaprice.org/book-reviews/thumbs-down/1903-food-rules.html [accessed: 1 March 2011].

Fallon Morell, S. 2010b. *The Omnivore's Dilemma by Michael Pollan* [Online: Weston A. Price Foundation]. Available at: http://www.westonaprice.org/book-reviews/thumbs-down/1822-the-omnivores-dilemma.html [accessed: 1 March 2011].

Fallon Morell, S. 2011. *Introductory remarks. A Critique of the 2010 USDA Dietary Guidelines* [Online: Weston A. Price Foundation]. Available at: http://www.westonaprice.org/press/2133-press-conference-critique-of-the-2010-dietary-guidelines.html [accessed: 1 March 2011].

Foucault, M. 1990. *The History of Sexuality, Vol. 1: An Introduction.* New York: Random House.

Heldke, L.M. 2003. *Exotic Appetites: Ruminations of a Food Adventurer.* New York: Routledge.

hooks, B. 1992. *Black Looks: Race and Representation.* Cambridge: South End Press.

Jackson, P. 2010. Food stories: consumption in an age of anxiety. *Cultural Geographies*, 17(2), 147–65.

Jeffery, C. 2009. Michael Pollan Fixes Dinner (Extended Interview). *Mother Jones* [Online, March/April]. Available at: http://motherjones.com/print/21935 [accessed: 1 March 2011].

Levenstein, H. 2006. *Paradox of Plenty: A Social History of Eating in Modern America.* 2nd Edition. Berkeley: University of California Press.

Massey, D. 2005. *For Space.* London: Sage.

Masterjohn, C. 2005. The truth about the China Study. [Online: Cholesterol-and-Health.com] Available at: http://www.cholesterol-and-health.com/China-Study.html [accessed: 1 March 2011].

Masterjohn, C. 2007. Response to T. Colin Campbell. [Online: Cholesterol-and-Health.com] Available at: http://www.cholesterol-and-health.com/Campbell-Masterjohn.html [accessed: 1 March 2011].

Meghani, Z. and Kuzma. J. 2011. The "revolving door" between regulatory agencies and industry: a problem that requires reconceptualizing objectivity. *Journal of Agricultural and Environmental Ethics*, 24(6), 575–599.

Milburn, M.P. 2004. Indigenous nutrition: using traditional food knowledge to solve contemporary health problems. *The American Indian Quarterly*, 28(3), 411–434.

Pollan, M. 2006. *The Omnivore's Dilemma: A Natural History of Four Meals.* New York: The Penguin Press.

Pollan, M. 2008. *In Defense of Food: An Eater's Manifesto*. New York: The Penguin Press.

Pollan, M. 2009. *Food Rules: An Eater's Manual*. New York: The Penguin Press.

Price, W.A. 1939. *Nutrition and Physical Degeneration*. Lemon Grove, CA: Price-Pottenger Foundation.

Probyn, E. 2011. Indigestion of identities. *M/C: A Journal of Media and Culture* [Online], 2(7). Available at: http://journal.media-culture.org.au/9910/indigestion.php [accessed: 1 March 2011].

Said, E.W. 1978. *Orientalism: Western Conceptions of the Orient*. London: Routledge and Kegan Paul Ltd.

Scrinis, G. 2008. On the ideology of nutritionism. *Gastronomica*, 8(1), 39–48.

Sontag, S. 1964. *Against Interpretation, and Other Essays*. New York: Farrar, Straus & Giroux.

Stanley, J. 2010. How good is the evidence for the lipid hypothesis? *Lipid Technology*, 22(2), 39–41.

Taubes, G. 2007. *Good Calories, Bad Calories: Fats, Carbs, and the Controversial Science of Diet and Health*. New York: Anchor Books.

Torgovnick, M. 1990. *Gone Primitive: Savage Intellects, Modern Lives*. Chicago: University of Chicago Press.

USDA CNPP 2010. *Dietary guidelines for Americans* [Online: Center for Nutrition Policy and Promotion, United States Department of Agriculture]. Available at: http://www.cnpp.usda.gov/dietaryguidelines.htm [accessed: 1 March 2011].

Ussery, E. 2009. *In Defense of Food by Michael Pollan* [Online: Weston A. Price Foundation]. Available at: http://www.westonaprice.org/book-reviews/thumbs-down/1536-in-defense-of-food.html [accessed: 1 March 2011].

Weston A. Price Foundation. 2000. *About the Foundation.* [Online: Weston A. Price Foundation]. Available at: http://westonaprice.org/about-the-foundation [accessed: 1 March 2011].

Weston A. Price Foundation. 2011. *Weston A. Price Foundation Funding* [Online: Weston A. Price Foundation]. Available at: http://westonaprice.org/funding-3.html [accessed: 1 March 2011].

Willett, W.C. 2001. *Eat, Drink, and Be Healthy: the Harvard Medical School Guide to Healthy Eating*. New York: Simon & Schuster Source.

Wilson, A.C. 2004. Introduction: indigenous knowledge recovery is indigenous empowerment. *The American Indian Quarterly*, 28(3), 359–372.

# Thematic Tabs for Chapter 9

Colonial

Discourse

Emotion

Race

Science

Women

Chapter 9

# Feminist Nutrition:
# Difference, Decolonization, and
# Dietary Change

Allison and Jessica Hayes-Conroy[1]

**Editors' Note:** This chapter engages the themes of *colonial, discourse, emotion, race, science, and women.* The chapter contributes to the project of doing nutrition differently by offering a vision of feminist nutrition that exemplifies the potential parallels between feminist activism and nutrition promotion. Drawing upon the theme of decolonization particularly, the authors move from a critique of white/Western "core" nutrition values to an articulated vision of how nutrition might be rethought, and repracticed, by way of feminist lessons on difference, discourse, decentering, and desire. The authors conclude by offering some brief examples of how feminist nutrition has been operationalized.

## Why Feminist Nutrition?

As feminist geographers and food scholars, we have spent a lot of time imagining how feminism could benefit the perception and practice of nutrition, both in the United States and abroad. These imaginings have arisen not just from varied experiences in the sites of our research, which are multiple as well as virtual, but also from our daily lives, where we continually communicate and interact with others about food, health, and bodies. Our imaginings have also arisen through our experiences teaching about food – through classroom dialogue that attempts to uncover the complexities of what it takes to promote healthy, nourished bodies in an incredibly uneven social world. We have continually returned to feminist theory and practice in our visions for how to "do nutrition differently." This chapter lays out the reasons why we see feminism(s)[2] as particularly useful for such an endeavor.

In articulating the usefulness of feminist perspectives for nutrition, we wish to focus on how feminist scholars and activists have confronted the issue of

---

1   The authors have contributed equally to the writing and editing tasks within, and have chosen this name format to reflect their equal contributions.

2   Feminisms (with an "s") refers to the multiplicity of feminist perspectives – the fact that there is not just one, unified feminist perspective. We use the term "feminism" in this chapter to denote a project that is, at its core, multiplicitous and (sometimes) contradictory.

social difference, broadly, and more specifically, the task of promoting social justice through a postcolonial/decolonial[3] approach. The reason for this emphasis is, quite simply, that it is an area of focus that has been tremendously central to feminist organizing and thinking, but almost entirely untheorized in terms of popular nutrition (i.e. nutrition not just as a scientific and medical endeavor, but as a widespread perception and practice). Indeed, the feminist movement has developed and been transformed through a series of critiques that have questioned the universality of the movement's supposed "core" – a core that, many have argued, has been dominated by white, Western ideas of womanhood. As a result of these critiques, feminism's "core" is now, arguably, more multiplicitous – defined not by any one set agenda or Truth (save the notion that difference matters) but instead by an insistence that contradiction and conflict are an inevitable and important part of progressive feminist movement. At the same time, attention to power *beyond* patriarchy has become central to feminism, lest we not forget who has had – and, indeed, continues to have – the authority to define such "universal" truths and "core" values.

Notably, postcolonial feminism (sometimes also known as Third World or transnational feminism) has taken particular issue with this notion of universality, arguing that white, Western feminists have a history of misrepresenting non-Western women in an effort to claim global sisterhood. Black and Chicana feminists in the US have similarly argued that such "core" feminism has ignored their voices and concerns. In both cases, white/Western feminism has been accused of perpetuating the effects of colonialism by silencing and marginalizing different voices. These critiques are long standing, but also ongoing. As a result, many have called for efforts to "decenter" the center of feminism – that is, to recognize the partiality of supposed (white/Western) "universals" and to destabilize the hegemony of the assumed "core" (see, for example, Narayan and Harding 2000).

Core values and universal truths are no stranger to nutrition. Despite the obvious political, social, and cultural influences on peoples' perceptions and practice of nutrition, "healthy eating" in our (Western) society continues to be defined in universal terms, with a core set of oft-repeated values and assumptions. This is so despite the fact that there is also an almost limitless supply of conflicting claims regarding what is "best" to eat – for at the center of each claimant's argument is the notion that their way is *the* best way. As a particularly powerful claimant, nutrition science contends to (still) have a hold on *the best way* to eat by "keeping it simple." Indeed, many nutrition and public health professionals argue that:

> The answer to the question *"What should I eat?"* is actually pretty simple. (Harvard School of Public Health website, 2012 emphasis in original)

---

3   There are discussions elsewhere of the distinction between postcolonial and decolonial approaches. These discussions are important, but are beyond the scope of our chapter. For further reading see Mignolo 2007, pp. 163-164.

Eschewing all the conflicts of opinion that are born through politics and culture, as well as science, many public health professionals, following this narrative of simplicity, urge the public to rely on a core set of basic, dietary guidelines. Just about anyone can recount these. Eat a lot of vegetables, fruits, and whole grains; enjoy healthy fats (e.g. olive oil, fish, nuts), but avoid saturated fat (butter, red meat); drink water; limit salt and sugar; exercise and take in an appropriate amount of calories for your body and lifestyle (Harvard 2012). That these core values have changed little in 30 years is taken as evidence of their universal veracity and acceptability. And, even though we have recently moved from pyramid to plate (myplate.gov 2012), the central message remains the same – eat like we say and you will be healthy.

Our primary critique of these core values is not focused purely on the metabolic benefits of the guidelines themselves (though a critique of this nature is also in order), but rather, at a more meta-level, on what such a claim of simplicity assumes, and what it avoids. By "keeping it simple," public health advocates assume that the general public cannot handle complexity – that difference and discrepancy must be bracketed away in favor of quick, easy answers to the question of "what to eat?". Moreover, this assumption is further complicated by the fact that, in the US at least, many of the "target populations" for nutrition intervention are communities of color – particularly, Black, Latino, and Native American – while those who are charged with defining and disseminating the "simple" nutrition facts continue to operate from a white/Western social location. Thus, we have a situation where "core" (read: white/Western) nutrition facts are simplified (read: dumbed down) for a largely non-white target population, who are assumed to not be smart enough to keep themselves healthy (or so we are to surmise, judging by the rates of dietary disease in non-white communities, combined with the continual emphasis on education-as-cure). To be fair, issues of food *access* and *cost* have moved into the national spotlight in recent years (for example, through Michelle Obama's Let's Move campaign (Let's Move 2012), and the Obama Administration's Healthy Food Financing Initiative; HHS 2012). Yet, most nutrition intervention initiatives tend to take an "if only they knew" approach (Guthman 2008), focusing on educating individual consumers – presumed to be either uneducated or lazy in regard to the practice of "core" nutrition values – rather than, say, advocating for social justice or trying to ameliorate economic inequality.

We should clarify that we are not interested in singling out nutrition science here, as the problems extend far beyond the scientific. Certainly many in the world of foodie-ism – activists, academics, and authors – insist on similar "core" educational strategies. From Michael Pollan's *Food Rules* (2009) and Mark Bittman's *Food Matters* (2009) to Ann Cooper and Holmes' *Lunch Lessons* (2006) and Marion Nestle's *What to Eat* (2007), core values abound in the world of alternative, local, home cooked, organic, fresh, raw, and otherwise like-your-grandma-made-it food advocates. Indeed, one does not even have to turn to authors for such examples. In our own everyday conversations with students, colleagues, family and friends, we have been continually struck by how frequently universality is assumed; in our own (progressive) circles, dietary "truths" are thrown around as though nutritional correctness were as sure as gravity.

Yet, we are not arguing that any sort of firm "stand taking" against the (corporate, unhealthy, inequitable, inhumane) food system is worthy of harsh critique. The point is, rather, that such "stand taking" becomes dangerous when it is touted as universally salient, apolitical, and socio-culturally neutral – or, to put it differently, it becomes dangerous when it eschews difference and discrepancy in favor of simple, core answers. For example, we can see the dangerous effects of this universal approach in the way that "culturally appropriate" nutrition has been operationalized to pay superficial attention to culinary and linguistic differences, while still holding fast to the supremacy of core nutrition facts. Thus, food pyramids and plates are translated into different languages and culinary cultures, but the "facts" remain secure. In Oldway's Latin American Diet Pyramid, for example (Oldways 2009), a can of frijoles and a papaya signal "Latin" culinary tradition, and dancing has replaced jogging as a depiction of physical activity. Nowhere in this schema is there room for any real destabilization of the core – for a decentering of the center. Instead, the tokenism of Latin food and dance here actually serve to *strengthen* the legitimacy of the core, by suggesting that the USDA food pyramid is universally applicable and translatable.

Another problem with relying on a core set of nutrition facts is that it also disqualifies the possibility that nutrition might actually be "done" differently, and still done beneficially. If nutrition is fundamentally about a list of do's and don'ts, then the proposed solutions to dietary problems will necessarily also be limited and simple. For example, a quick search of nutrition-related grant opportunities will illustrate that most funding agencies are interested in funding research on how to best get the word out, assuming again that it is a lack of ("core" value) education that is the root cause of dietary disease. As our pyramids are replaced by plates in an effort to *further* simplify the tenets of healthy eating, we might want to start asking ourselves whether more simplicity is really what we need.

### What does Feminist Nutrition look like?

Building the case for why we need a *feminist* approach to nutrition could take much longer. But at some point it's important to stop critiquing what's out there and start imagining what could be. If, as the critique above suggests, it's not *simplicity* we need, then how are we to build *complexity* into a perception and practice of nutrition that still is able to achieve concrete results? How can we ensure the promotion of 'healthier' people, eating 'healthier' foods, while holding in question how 'healthy' is best defined and measured? While this is undeniably a tall task, what is so exciting and promising about feminism is its legacy of handling complexity and contradiction in effective and productive ways. Indeed, the feminist movement has had much practice in turning impasse into inspiration, difference into dialogue, and complexity into consciousness-raising. While there are many ways to imagine how feminism could transform the perception and practice of nutrition, here we wish to focus briefly on four key themes that we think are central to crafting a "Feminist Nutrition." Those are: 1. Women and Gender, 2. Standpoints and Situated Knowledge, 3. Decentering and Decolonization, and 4. Im(purity) and Embodied Process.

## 1 Women and Gender: Recognizing Difference

It might be surprising to some readers that we have not yet explicitly discussed women, or gender, in an article on Feminist Nutrition – at least not in direct relation to food. This is purely functional. Above, we have chosen to focus our analysis of nutrition in terms of a critique of "core" (white/Western/scientific/universal) values. As we have suggested, because feminism has faced and grappled with similar critiques, the lessons learned from such struggles in the feminist movement could help to improve the perception and practice of nutrition as well. In this sense, the parallel that we are drawing above is between the trajectories or movements of feminism and nutrition promotion, movements which have intended to better the lives of individuals and groups in need. Nevertheless, just as the feminist movement has *specifically* sought to improve lives by promoting gender equality, in connection with race, class, sexuality, and other forms of social difference, there are also reasons why nutrition would similarly benefit from a *specific* focus on issues of women and gender. Indeed, although we have chosen to focus above on other aspects of feminist theory and practice, there are certainly rich and deep connections between women/gender and nutrition/healthy eating; so many so, in fact, that it would be impossible to discuss them all here. But let us mention a few.

Feminist food studies scholars have long argued that women play a unique part in the food system, from their role in the processes of agricultural production to the parts they play in the processes of food consumption. Women make up over fifty percent of all agricultural workers globally, but they often do not own their own land, or make their own money – and thus are more liable to go hungry (Seager 2008; Allen and Sachs 2007; Barndt 1999). They are likely the food provisioners for their families, as well as the means through which culinary-based cultural reproduction happens in their homes and communities, and yet many women also spend much time serving others in food-based retail and service jobs (Avankian and Habar 2005; Counihan 2009; Allen and Sachs 2007). And, women also struggle in unique ways with issues of dieting, body image, and weight gain, including struggles that lead to exceptionally high rates of disordered eating among women (Bordo 1993; Thompson 1996). In short, all of this means that gender is a significant lens through which we can begin to understand how social differences come to matter to people's relationships to food and nutrition. By studying gendered difference, as the feminist movement has illustrated, we can begin to make visible and to question the unspoken assumptions of the supposed "core." Importantly, if we are careful with this critique, it should also lead us to investigations of other, (always) interconnected forms of social difference and inequality – e.g. those based on race, class, sexuality, religion, age, ability, colonial status, etc.

But, the goal of exposing such inequality and strife is not to offer up yet another singular way of eating *right*. Rather, taken as a whole, the above scholarship is most important as a study in complexity and contradiction. It highlights, for example, the socio-cultural and political intricacies of food decision-making, the (often) ambivalent relationship that women have with food-based gendered roles, the ways that women have struggled and learned to make do, and also the ways that food has become a source of empowerment for many along the way. These stories don't represent complete answers to the problems of our food system, and some may even stand at odds with each other, morally speaking, but they are powerful in their partiality and contextuality. They suggest that unique stories *matter* to the practice of nutrition. They signal – forcefully, starkly – that a universal answer to the question "what's good to eat?" is both academically sloppy and socially unjust. And, they insist instead that we must focus our efforts on ensuring that the nutrition advice that we seek to offer be emergent and negotiated, as well as frequently reassessed.

## 2 Standpoints and Situated Knowledge: An Objective Approach

In seeking to specify a way to "do nutrition differently," we might be concerned about questions of veracity and rigor. These are important concerns. If nutrition advice is to be (only) emergent and negotiated, as we suggest above, does that mean that all advice is relative? And, if this is the case, how are we to make significant progress on our path towards 'healthier' people and communities? For similar reasons of effectiveness, it has been crucial in the feminist movement to specify what exactly we achieve when we insist upon hearing *different* voices. Feminist scholar Sandra Harding calls what we achieve "strong objectivity" (Harding 1993). Harding insists that, far from promoting more subjective accounts, by listening to those who are marginalized or oppressed by "the core," we are able to arrive at a *more* complete picture – one that is able to both account for marginalized experiences and expose the hidden assumptions of the (ostensibly neutral) center. It is for this reason that feminists like Harding have promoted the concept of privileged "standpoints" – social locations from which it becomes possible to provide more objective, rigorous accounts (Hesse-Biber and Leavy 2007). As feminist standpoint scholars insist, storytelling is a necessary part of effective feminist practice. In terms of nutrition, this means that the details, differences, and discrepancies that people's life experiences reveal play a crucial role in producing effective, meaningful advice.

Standpoint scholars have also warned, however, that occupying a social location *outside* the core does not necessarily ensure critical thought and action. At the same time, occupying a more central/core social position does not necessarily preclude someone from participating effectively in the feminist movement. This is because strong objectivity also depends upon the ability for different standpoints to come together in democratic dialogue – to work together in solidarity to think and act critically. Of course, this is more easily said than done. Because ideas and actions emerging from the core often carry more social legitimacy (e.g. scientific truth claims), the power differential between various social locations can often prevent effective, critical dialogue. An important viewpoint to help equalize different knowledge claims is the notion of "situated knowledges" (Haraway 1988). Biologist and feminist scholar Donna Haraway coined the phrase in 1988 in an effort to shed more light on the meaning and practice of objectivity in science and in feminism. Haraway offers that objectivity is *embodied* – that it emerges out of particular contexts, places, and peoples – and as such that it necessarily accommodates paradox and critique. She insists on the particularity and embodied nature of *all* vision, such that "objectivity turns out to be about particular and specific embodiment and definitely not about the false vision promising transcendence of all limits and responsibility" (1988, 582). This definition of objectivity is particularly important for scientific claims (such as those of nutrition science) because it simultaneously offers science a space in the dialogue, while also insisting that such perspectives are no less partial than any others. She continues:

> The moral is simple: only partial perspective promises objective vision. All Western cultural narratives about objectivity are allegories of the ideologies governing the relations of what we call mind and body, distance and responsibility. Feminist objectivity is about limited location and situated knowledge, not about transcendence and splitting of subject and object. It allows us to become answerable for *what we learn how to see* (1988, 583 emphasis added).

In terms of the practice of nutrition, then, the concepts of strong objectivity and situated knowledge suggest that we need to spend more time attending to the limits and location of what we know about "what's good to eat." In addition, by attending to standpoints of the oppressed and silenced particularly, we have a better chance of recognizing and reaffirming the *partiality* of core perspectives. But, regardless of our standpoint, to learn the skills needed to see food more fully, and to imagine new ways of practicing nutrition, we need to share the struggle. Following Haraway's vision (1988), the practice of nutrition would necessarily involve activities aimed at both the collective contestation and deconstruction of dominant perceptions and practices, as well as the building of opportunities for sharing partial knowledges and ways of doing. Thus, rather than a distinct and pre-determined set of core values and goals, nutrition becomes a collective, negotiated process of transformation towards healthier people and communities.

*3 Decentering and Decolonization: Origin Stories*

We began our critique of popular nutrition by discussing the problems with core assumptions, and particularly the colonialist, missionary underpinnings of an "if they only knew" approach (Guthman 2008). Dealing effectively with such deeply embedded and widespread power inequities *in practice* is surely not an easy task. But, methods and pedagogies aimed at destabilizing the core are also no stranger to feminist practice, nor to the healthy/alternative food movements[4]. Indeed, currently, at the intersection of activist and academic work within the alternative agri-food community, we have seen a marked emergence of interest and involvement in what is often termed "dietary decolonization." Dietary decolonization appears to encompass a range of agri-food practices, among a range of peoples, including efforts in food system re-localization, veganism, and pre-contact food challenges, among African-American, Latina, Native American and Hawaiian groups. And women appear to often be leaders of these activities, perhaps for the same reasons that they are often leaders in many other food-based social movements: concern for the health of their families, communities, and ecologies. What connects these varied initiatives is an attempt to name, specify and put into practice a decolonized approach to healthy eating. In some ways, this trend mirrors the work of scholar activists like Vandana Shiva, whose work is focused on agriculture and health issues in the global South; but now, and increasingly, this call for nutritional decolonization is also taking place in the global North.

In addition, and importantly for understanding the *practice* of decolonization, this trend is also increasingly identified as taking place at the scale of the body. Thus, not only is there an emphasis on connecting personal food choices to political economic inequities, there is also an impulse to see the body as a strategic location for social change. Decolonizing therefore becomes a practice that must occur at the level of taste and desire. One example of a decolonial food project is Breeze Harper's (2010) edited volume, *Sistah Vegan: Black Female Vegans Speak on Food, Identity, Health and Society*. The book is a collection of essays, poems and stories that recount a project of claiming "veganism as health activism that resists institutionalized racism and neocolonialism" (jones 2010, 187). More specifically,

> [The book] presents veganism as a Black feminist and antiracist practice... [and] illuminates inconvenient connections that the feminist, antiracist, animal liberation, and environmental movements have too long ignored.... [it] demonstrates the necessity of decolonizing desire, not only among formerly colonized peoples but among all of us whose socially constructed appetites are eating up the world (jones 2010, 187-188).

Such projects are without a doubt strongly linked to important questions in both feminism and postcolonialism – not only by articulating a gendered approach to questions of food imperialism, but also by exemplifying the political importance

---

4  While we recognize that advocating for healthy food and advocating for alternative food are not *always* the same thing, we collapse them here because the two movements so frequently intertwine, reinforce, and legitimize each other that it is misleading to speak of them as wholly separate forces.

of the material, corporeal body as a strategic location to do such decolonial work. The important and difficult question of *how* the material body can (and cannot) accomplish such work is the topic of the next section, below. First, however, it is important for us to specify more specifically *why* these projects are particularly radical, critical, and feminist. After all, cultivating a desire for vegan food (to continue with the current example) is arguably a fairly mainstream food practice (within significant links to corporate agribusiness, for instance).

Without discounting the significance of veganism, we want to propose that such decolonial food projects are *most* groundbreaking in the lessons (and certainly the bodily inspiration) that they offer for decentering whiteness – that is, for "decentering the center" (Narayan and Harding 2000). An important recent critique of healthy/alternative food activism has focused on its role in the reproduction of a seemingly hegemonic whiteness (Slocum 2011; Alkon and Agyeman 2011; Guthman 2008; Hayes-Conroy and Hayes-Conroy 2010). As numerous food scholars have shown, slow food, school gardens, farm markets, and Whole Foods stores are frequently coded as white spaces (as well as socio-economically elite). Food scholars have drawn from nuanced understandings of race to articulate how whiteness becomes a dominant organizing force in such healthy food spaces, how it circulates, and also how it can be shifted. While these stories are certainly an important way to recognize and challenge white dominance and privilege, a focus on whiteness alone also threatens to reify the monolith of whiteness *as the only story out there*. Of course, we all know that there *are* many other stories. And food scholars have surely told some of these stories as well, specifying some of the ways that the whiteness of healthy/alternative food is disrupted through daily, food-body interactions (Slocum 2007; Harper 2010). But there is still more to say.

Postcolonial feminist practice stresses the importance of origin stories in the quest to decenter whiteness. For healthy/alternative food activism, this decentering is not accomplished by recognizing that food initiatives such as veganism can be "black too," or "latina too," but instead by realizing that *the very impulse or desire to engage* in food activism (veganism included) might be, quite expressly, *not* white. That is, the motivational drive to change one's food habits in response to healthy/alternative knowledge claims is not just pluralistically multi-racial, but it can and does also emerge from *specifically* non-white communities, contexts and experiences. In this way, decolonial food projects should remind us of activist work in the environmental justice movement, as well as the farmworker justice movement, which have also emerged largely from marginalized, non-white communities. Certainly the movements share some important genealogies that are quite distinctly *not* white. To be clear, this does not mean that we must deny that veganism as a culinary practice has *also* been racially coded as white. Certainly this story is an important part of the puzzle. But to label the origins of veganism as white-and-only-white (or even white-and-predominantly-white) is to deny the power of non-white vegans to inspire, influence and create.

More broadly, feminist postcolonial activism has already long since insisted upon the importance of such genealogies, or origin stories, from feminist activists of the global South. Indeed, in subverting the notion of a singular, "core" feminism, feminist postcolonial scholars have continually encouraged us to recognize the emergence of a feminist impulse from many distinct global contexts. As Chandra Mohanty reminds us,

> Decolonization has always been central to the project of Third World feminist theorizing – and much of [her] own work has been inspired by these particular feminist genealogies (Mohanty 8, 2003).

Thus, in regard to *doing nutrition differently*, postcolonial feminist practices remind us to pay increased attention to the origin stories of healthy/alternative food projects – that is, to the places and peoples from which healthy/alternative food impulses *originate*. As we recognize and tell these origin stories, and make them central to our perception and practice of nutrition, we can begin to de-center whiteness as the only story out there.

*4 Im(purity) and Embodied Process: Decolonizing Desire*

We have already mentioned how some feminist and decolonial food activists have begun to re-conceptualize the *material* body, including bodily affects/emotions, as playing an active and necessary role in food-based social change. But what does it actually mean to decolonize one's bodily tastes and desires? Elsewhere, we have specified the importance of the material body in the processes of food decision-making, including the decision of whether or not to engage in healthy/ alternative food activism (Hayes-Conroy and Hayes-Conroy 2008, 2010). More specifically, we have insisted that the material body can both confirm and disrupt social trends in eating habits based on gender, race, class, age and other forms of social difference, because the material body is both developmental *and* unpredictable. Thus, we have suggested that the material processes that produce tastes and other food preferences are best described as a rhizome of forces – some structural and some haphazard – that intersect to produce specific moments of food-body interaction. In short, following in the footsteps of a number of feminist scholars interested in food and bodily materiality (McWhorter 1999; Probyn 2000; Roe 2006; Colls 2007; Slocum 2008; Longhurst et al. 2009), we have insisted that bodies *matter* – quite literally – to any project of food-based social change.

More specifically, however, we also have wanted to highlight that bodies often come to matter in quite *different* ways. For example, to some, the "core" values of white/Western nutrition might be inspiring, and the foods that correspond to and confirm these values might feel familiar and comforting. But, to others, these core values may have a more "chilling" effect (Guthman 2008), and the related foods may taste unfamiliar, or even disgusting. While attentiveness to social differences like gender, race, class, and age can help us to speculate about how different groups *might* respond to particular foods or nutritional concepts, bodies (and the uneven socio-material systems in which they develop) are much too complicated to allow for any fixed, easy predictions. Indeed, as so many feminist scholars have shown, the body consistently thwarts scholarly attempts at oversimplification or categorization.

Given all this, it is important to recognize that nutrition projects cannot succeed when they simply attempt to target and re-train bodies that don't prefer the "correct" foods. This sort of disciplinary approach to bodies is, arguably, not that different from colonial attempts to shame women into "correct" performances of gender and sexuality, or to frighten pantheists into Christian worship. Thus, while we must recognize that material bodies *matter* in the process of social transformation, we must also understand that our goal is not, and cannot be, the production of increasingly *similar* bodily preferences and desires. This is true not only in regard to the projects of the supposed nutritional "core," but also in regard to projects of the periphery that attempt to actively subvert this core.

Indeed, as postcolonial feminism reminds us, in our quest to de-center the center (through both storytelling and bodily practice), purity and coherence are *not* what we are seeking. That is, decolonial projects are not primarily about identifying a pure, uncolonized time or body to which people could or should return. And they are not about identifying a coherent, pre-colonial group of others that exist, and eat, in opposition to the colonizer. Indeed, as Jinga Desai (et al. 2010) tell us, we need to "put into question [any such] an approach to difference that takes 'accuracy' as the measure of its success." In other words, the search for a purely coherent decolonized subject is, while perhaps a motivational goal for some, not a realistic requirement or even an appropriate focus for most embodied, decolonial work.

As the authors go on to explain:

> For us, the [decentering/decolonial] project … is not about getting beyond, per se, and can never be complete, as it both succeeds and fails continually. In other words, its project cannot be decided upon once and for all by a description of how things "really are," which would close down further inquiry into how objects and subjects of knowledge are brought into being. Pedagogy itself constantly reminds us that critical readings are always necessary and need to be redone, relearned, and rewritten and that knowledge and its production are in a constant state of being contested, analyzed and reformulated. (Desai et al. 2010, 55)

Taken as a statement about the material body and its *embodied* knowledge, the above quotation suggests that the project of bodily decolonization is never complete, and never without contradiction. It reminds us that, as we attend to the monoliths of domination (the power of the "core"), we must also allow for material, bodily movement. If we do not want to reproduce the distinctions and inequities that we wish to disrupt, we need to recognize that contradiction and complexity are central organizing features of both individual, bodily subjectivity and collective action. In *Decolonizing Methodologies*, Linda Tuhiwai Smith, notes that:

One of the many criticisms that gets leveled at indigenous intellectuals or activists is that our Western education precludes us from writing or speaking from a 'real' and authentic indigenous position. Of course, those who do speak from a more 'traditional' indigenous point of view are criticized because they do not make sense… (1999, 14).

Connecting this back to food and nutrition, Smith's comments remind us that we should not look to produce pure, authentically "decolonial" bodies. Not only is this search futile, but also, by focusing on the purity of the distinction between colonizer and colonized, such an attempt actually strengthens the core – by allowing the monolith of whiteness to become the whole story, and a purely decolonized other to become the antidote.

We began this chapter by suggesting that we need to allow for more complexity and contradiction in our perception and practice of nutrition. This last section has suggested that part of this allowance means recognizing the body itself as a complex and often contradictory actor in the process of food decision-making. After all, decolonization is not a pill that you can take that will wipe out all traces of the colonizer's tastes and desires. And indeed, were it even possible to take such a pill, decolonization could never be "complete" because our bodies do not exist in social (or biological) vacuums. Instead, as the above scholarship tells us, decolonization is necessarily a more partial and negotiated process that strives for neither authenticity nor purity, but rather the clarity of situated, embodied knowledge and the certainty of material change.

## Operationalizing: Concluding Thoughts

If we want to discuss how to operationalize the imagined perceptions and practices that we describe above, we have to first ask who is to be engaged in this feminist nutrition? We could say "everybody(!)," and run the risk of romanticizing unity and assuming inclusivity, when neither are clearly assured. Or, we could say, "nutrition practitioners," and single out one (presumably) problematic group, when we have already said that the problem is widespread and pervasive. Maybe then the best answer to the question of operationalization, like our vision above, is that it is *also* necessarily situated and contextual. As a vision for how to *approach* nutrition, operationalization depends on who is approaching, where, and why.

There are indeed already numerous practitioners and activists who exemplify and operationalize some of the ideas that we discuss above. Patricia Williams, for example, is an applied human nutritionist from Nova Scotia who utilizes a participatory action approach in her work on hunger and food security. Williams utilizes the methods of storytelling and participatory food costing, among others, to give voice to women who struggle with hunger, and to make visible the socio-economic inequities that exist behind individual food choice. Her methods are meant to reduce the potential for unjustified "assumptions and stereotypes," and to build a more inclusive model that promotes a community's capacity to affect change (PARTC 2012). Similarly, Farm Fresh Choice, the Ecology Center's food justice program in Berkeley, CA, trains adult mentors and teen leaders to facilitate workshops that use a variety of culturally embedded origin stories to develop food and nutrition knowledge. These workshops approach health holistically, but also in ways that reaffirm non-white genealogies of nourishment, empowerment, and healing (Ecology Center 2012). Both Williams' work and the work of Farm Fresh Choice illustrate interesting and effective ways to begin to operationalize Feminist Nutrition.

Finally, in closing, we also want to reassert the place of science at the table of feminist nutrition. To be sure, science can and does help to ensure radical, critical thinking, even as it can also be co-opted by forces that reinforce hierarchy and undermine diversity. Particularly, we are thinking of the ways in which science has been utilized in attempts to bring justice to communities exposed to toxic waste, heavy metals and other forms of environmental pollution (Harrison 2011, Allen 2003, Coburn 2005). We see that these issues are becoming increasingly important to nutritional health as well, especially as Bisphenol A (BPA) and other endocrine disrupting chemicals are found to have potentially harmful effects on metabolic processes (Kristof 2012, Guthman 2011). These interdisciplinary studies can help to not only explain high rates of obesity, among many other health 'epidemics' of late, but can also work to contextualize issues like obesity within the larger, political economic contexts in which such 'individual' conditions are produced. This contextualization certainly encourages us all to rethink education-as-cure models of public health intervention, and instead to support continued scientific research that highlights the relationship between human health disparities and political economic inequity. But, perhaps equally as radical as this, science also has the potential to shed new light on the social nature of bodily material *itself*, including, for example, the socio-materiality of processes like taste or digestion. We need scholars of science, in dialogue with many other critical thinkers, in order to make sense of the messy jumble of social and metabolic exchanges through which bodies come to act and react *differently* to food, producing health outcomes that are far from predictable. Indeed, it is scientific inquiry, in combination with political economic analysis and social critique, that can help us to truly see the situated contexts from which we, as minded bodies, come to *know* what is good to eat.

## References

Allen, Barbara L. 2003. *Uneasy Alchemy Citizens and Experts in Louisiana's Chemical Corridor Disputes*. Cambridge: MIT Press.

Allen, Patricia and Carolyn Sachs. 2007. Women and Food Chains: The Gendered Politics of Food. *International Journal of Sociology of Food and Agriculture.* 15: 1.

Alkon, Alison and Julian Agyeman. 2011. *Cultivating Food Justice: Race, Class and Sustainability.* Cambridge: MIT Press.

Avakian, Arlene V. and Barbara Habar. 2005. *From Betty Crocker to Feminist Food Studies: Critical Perspectives on Women and Food.* Amherst: University of Massachusetts Press.

Barndt, Deborah. 1999. *Women Working the NAFTA Food Chain: Women, Food and Globalization.* Toronto: Sumach Press.

Bittman, Mark. 2009. *Food Matters: A Guide to Conscious Eating with More Than 75 Recipes.* New York: Simon & Schuster.

Bordo, Susan. 1993. *Unbearable Weight: Feminism, Western Culture, and the Body.* Berkeley: University of California Press.

Coburn, Jason. 2005. *Street Science: Community Knowledge and Environmental Health Justice.* Cambridge: The MIT Press.

Colls, Rachel. 2007. Materialising Bodily Matter: Intra-action and the Embodiment of 'Fat'. *Geoforum.* 38: 353-365.

Cooper, Ann and Lisa Holmes. 2006. *Lunch Lessons: Changing the Way We Feed Our Children.* New York: William Morrow.

Counihan, Carole. 2009. *A Tortilla Is Like Life: Food and Culture in the San Luis Valley of Colorado.* Austin: University of Texas Press.

Desai, Jinga, Danielle Bouchard and Diane Detournay. 2010. "Disavowed Legacies and Honorable Thievery: The Work of the "Transnational" in Feminist and LGBTQ Studies." In Amanda Lock Swarr and Richa Nagar, Eds. *Critical Transnational Feminist Practice.* SUNY Press.

Ecology Center. 2012. Farm Fresh Choice. Online at www.ecologycenter.org/ffc. Cited September 2012.

Guthman, Julie. 2008. "'If They Only Knew': Color Blindness and Universalism in California Alternative Food Institutions. *The Professional Geographer.* 60: 3, 387-397.

Guthman, Julie. 2011. *Weighing In: Obesity, Food Justice, and the Limits of Capitalism.* Berkeley: University of California Press.

Haraway, Donna. 1988. "Situated Knowledges: The Science Question in Feminism and the Privilege of Partial Perspective." *Feminist Studies.* 14: 3, 575-599.

Harding, Sandra. 1993. "Rethinking Standpoint Epistemology: What Is "Strong Objectivity?" in Linda Alcoff and Elizabeth Potter, Eds. *Feminist Epistemologies.* New York: Routledge.

Harper, A. Breeze. 2010. *Sistah Vegan: Food, Identity, Health, and Society: Black Female Vegans Speak.* Brooklyn: Lantern Books.

Harrison, Jill. 2011. *Pesticide Drift and the Pursuit of Environmental Justice.* Cambridge: MIT Press.

Harvard School of Public Health. 2012. "The Nutrition Source: What Should I Eat?" Online at: www.hsph.harvard.edu/nutritionsource/what-should-you-eat. Cited September 2012.

Hayes-Conroy, Allison and Jessica Hayes-Conroy. 2008. Taking Back Taste: Feminism, Food and Visceral Politics. *Gender, Place and Culture.* 15: 5, 461-473.

Hayes-Conroy, Allison and Jessica Hayes-Conroy. 2010. Visceral Difference: Variation in Feeling (Slow) Food. *Environment and Planning A.* 42, 2956-2971.

Hesse-Biber, Sharlene and Patricia L. Leavy. 2007. *Feminist Research Practice: A Primer.* Thousand Oaks: Sage Publications.

HHS (US Department of Health and Human Services) 2012. Healthy Food Financing Initiative. Online at: www.acf.hhs.gov/programs/ocs/ocs_food. html. Cited September 2012.

jones, pattrice. 2010. "Afterword: Liberation as Connection and the Decolonization of Desire," in A. Breeze Harper. *Sistah Vegan: Food, Identity, Health, and Society: Black Female Vegans Speak.* Brooklyn: Lantern Books.

Kristof, Nicholas D. 2012. "Big Chem, Big Harm?" *New York Times,* Sunday Review: The Opinion Pages, August 25, 2012.

Let's Move. 2012. "Healthy Communities," Online at: www.letsmove.gov/ healthy-communities. Cited September 2012.

Longhurst, Robyn, Lynda Johnston, and Elsie Ho. 2009. A Visceral Approach: Cooking 'At Home' with Migrant Women in Hamilton, New Zealand. *Transactions of the Institute of British Geographers.* NS 34, 333-345.

McWhorter, Ladelle. 1999. *Bodies and Pleasures: Foucault and the Politics of Sexual Normalization.* Bloomington: Indiana University Press.

Mignolo, Walter D. 2007. Coloniality of Power and Decolonial Thinking. *Cultural Studies,* 21: 2-3, 155-167.

Mohanty, Chandra T. 2003. *Feminism Without Borders: Decolonizing Theory, Practicing Solidarity.* Durham: Duke University Press Books.

myplate.gov 2012. "USDA Choose My Plate (dot) gov" Online at: myplate.gov. Cited September 2012.

Narayan, Uma and Sandra Harding. 2000. *Decentering the Center: Philosophy for a Multicultural, Postcolonial, and Feminist World.* Bloomington: Indiana University Press.

Nestle, Marion. 2007. *What to Eat.* New York: North Point Press.

Oldways. 2009. "Latino Diet & Pyramid." Online at oldwayspt.org/resources/ heritage-pyramids/latino-diet-pyramid, Cited September 2012.

PARTC. 2012. "Participatory Action Research and Training Center on Food Security." Online at www.foodsecurityresearchcentre.ca. Cited September 2012.

Pollan, Michael. 2009. *Food Rules: An Eater's Manual.* New York: Penguin Books.

Probyn, Elspeth. 2000. *Carnal Appetites: FoodSexIdentities.* London: Routledge.

Roe, Emma J. 2006. Things becoming food and the embodied, material practices of an organic food consumer. *Sociologia Ruralis,* 46(2), 104-121.

Seager, Joni. 2008. *The Penguin Atlas of Women in the World: Fourth Edition.* New York: Penguin Books.

Slocum, R. 2007. Whiteness, space and alternative food practice. *Geoforum.* 38(3): 520-533.

Slocum, R. 2008. Thinking race through feminist corporeal theory: divisions and intimacies at the Minneapolis Farmers' Market. *Social and Cultural Geography.* 9(8): 849-869.

Slocum, Rachel. 2011. Race in the Study of Food. *Progress in Human Geography.* 35: 3, 303-327.

Smith, Linda T. 1999. *Decolonizing Methodologies: Research and Indigenous Peoples.* London: Zed Books.

Thompson, Becky. 1996. *A Hunger So Wide and So Deep: A Multiracial View of Women's Eating Problems.* Minneapolis: University of Minnesota.

# Thematic Tabs for Chapter 10

*Body*

*Emotion*

*Women*

# Chapter 10
# Nutrition is . . .

Laura Newcomer

**Editors' Note:** This chapter poetically intersects with three thematic tabs: *body, emotion* and *women*. Newcomer reflects on an all-too-common theme today, the dissolution of healthy food-body relationships, and demonstrates a profound interweaving of emotion, intellect, physical activity, trauma, hunger, and fatigue in her experiences of 'disordered' eating. Newcomer's words highlight the kind of complex and situated struggle that many women experience with respect to their food-body relationships.

## I.

Calories out. Out!
Calories in (so few of them!) and
expanding the deficit. Nothing matters more
than growing smaller. Bone-knees knock against each other on
the day's second walk home from the gym.
When I cross the street
I can barely lift my foot up and
over the sidewalk's edge; I cannot be sure that I will not fall
down, there, in the gutter.

## II.

Him feeding me
lies and me swallowing
them. Me thinking myself worthless
and his fingers
in my underwear though
I do not want them under there, though
I had tried— once, perhaps, now a long time ago— to say no.
He pushes me latkes, leering;
I push them around my plate. I push down
any part of me who believes I deserve anything
better
than him pushing me— into rough stone
walls, against his mauve Miata, down

into the mattress, his body wet-bucking, mine motionless,
dry.


**III.**

Eating less to survive. Eating less to avoid
seeing; to avoid feeling; so that I can become
"Normal." Or,
at least,
acceptable. Contained
body, constricted mind (it is immobilized
by counting). She is "so nice," now,
all of the time. She is so quiet. She gets good grades.
She is killing herself. People praise her.


**IV.**

Mother finally being proud. It feeds me.
But she does not brag about her daughter's body. It
is too skinny, she says; she can see bones poking through. She says those calves
look like a super model's calves. She says
That's not a good thing.


**V.**

I was told it was such a good thing.
I and every woman, told: If we could just
Look
Like
That.


**VI.**

Me embodying the hypocrisy.
I am what they would have me be
Still I am not good enough. I expose:
There is no
Good.
Enough.
Nutrition is

**I.**

Binging on protein bars at 11:30 at night. Six of them already;
my belly aches and still I cannot
stop. I am so hungry.


**II.**

Eating doughnuts for breakfast and sneaking more
doughnuts into my purse while my friends (who do and do not
know me, for they know not what I do) are not watching (oh friends, I am
so ashamed.) As I edit the newspaper, alone
in a small office overlooking the quad,
I will pull the old fashioned from its wax paper wrapper. Then the chocolate
glazed. Next finally
the lemon cream-filled, and halfway through I will know
how desperately I want to stop and even then
I will not be able to stop.
I am so hungry.


**III.**

Stomachaches and headaches and tiredness
and self-loathing and shame. I disgust myself. I hide.
I no longer pick up his calls. I do not see him and still
a voice tells me
I am worthless.
It is my own.
Nutrition is


**I.**

Working to lose it all again. Calories out, calories in,
and expanding the deficit. It is enough (isn't it?)
to drive you mad.
And how many more years of this?
Eight more years
of this.
Nutrition is

**I.**

Admitting that I am struggling.
Deciding, for reasons that I cannot now remember nor, at the time,
could perhaps fully comprehend, to do— something—
about it.
This is really fucking hard.


**II.**

Learning how to eat. Learning what my body likes
and does not like. Learning to respect
those preferences. Learning to recognize
my own hunger
and whether I am eating
out of avoidance or fear. Learning to say Yes— to full fat, to bigger portions, to
those foods whose names I do not know or for which
I have not memorized the exact caloric count, to the experience
of pleasure—and
just as importantly,
to say No.


**III.**

Saying no to his fucking fingers,
fingers fucking don't just go where they are not
welcomed in. Hands take them there. A body
has been violated. I am beginning to see.


**IV.**

Learning to see myself. Seeing that
I Am is different than
He Says, than They Say, than Mother Said, different even
than "so says a voice inside of me."
Learning that I am separate from all of that. I stand distinct,
an autonomous being. I exist.
I have a say. I am allowed to make decisions
about the treatment of this body. My body. I am a person
worthy of love. I am beginning to love me.

### III.

Learning how to love other people, not by
sacrificing the marrow from out of me.
I want so badly for the world to be full—oh, to bursting!—
with nurtured people. Lest I play a zero-sum game,
that number must include me.

### IV.

Learning how to feed myself:
Writing affirmations. Working to believe in myself. Taking
long walks. Taking a break from the gym. Singing
again. Singing even in front of other people.
Playing the guitar. Making lists of
things that I love. Reading authors whose writing makes
my chest ache. In a good way. Yelling. Being loud.
Being more honest. Sharing with people who are close to me and
who have earned the right to hear. Working to remove poisons
from my life— people, relationships, thoughts, deeds. Doing yoga.
Lying on my back on my bedroom rug, listening to my favorite musicians
play. Sitting in the woods. Climbing trees.
Teaching myself how to can peaches. Re-learning how to sew. Painting in my bare
feet
in the grass beside my best friend. Designing collages. Kickboxing.
Working a little less. Socializing more. Breathing:
long, slow, deliberate. Dancing. Having fun.
Writing poetry again.
Nutrition is

### I.

On some days, eating organic oatmeal for breakfast,
grilled salmon and salad for lunch,
homemade lentil soup for dinner. Some days eating
fistfuls of cashews for breakfast,
snacking all the way to lunch, swallowing chocolate
and Mentos 'til a dinner of greasy Chinese. Some days
exactly what I need is Mentos and Chinese.
Every day trying
to listen to my body. Many days
failing to do so. Some days

getting it just right.

## II.

Re-teaching myself how to eat
on a regular basis. Thinking that I've got it down
and then overeating for one, maybe twenty-five
days. Thinking that I've got it down and then not eating
enough and exercising too much for too many days in a row. Getting frustrated
because I thought I had it down. Trying to accept
that I will never have it— or anything—
completely down; this is a process;
sometimes I will feel frustrated;
sometimes I will feel fabulous;
sometimes I will tear my own self apart and sometimes
I will love my stomach even as it aches.

## III.

The authority to define and re-define
what "just right" means, on a daily basis, for my body and
for me. Listening. Trying my best to be
mindful. Working to see every body
beautiful. Trying my best to approach other people
and myself
with compassion
and love. That's the nutrient.

# Thematic Tabs for Chapter 11

> *Colonial*

> *Nature*

> *Race*

> *Science*

> *Structure*

> *Women*

Resource Tabs for Chapter 11

## Chapter 11

# Another Way of Doing Health:
# Lessons from the Zapatista Autonomous
# Communities in Chiapas, Mexico

### Chris Rodriguez

The most precious thing you can give to a movement is health.
— Zapatista promotores de salud, Caracol Oventic[1]

**Editors' Note:** This chapter connects to the thematic tabs *colonial, nature, race, science, structure* and *women*. As a chef, scholar and activist advocating decolonization, Rodriguez reads the autonomous health and nutrition projects of the Zapatistas as decolonial practices—actions aimed at de-linking from the intersectional hierarchies created by the colonial system. Rodriguez's contribution is to demonstrate how matters of nutrition and health intermingle with the Zapatista struggle over their ancestral territories.

A nutritional health perspective rooted in indigenous principles brings to the center what modern nutrition and progressive-liberal fronts of food justice activism ignore, or worse yet, work to erase—the decolonial imperative. Grassroots indigenous social movements like the Zapatista Autonomous Communities in Chiapas, Mexico working to end hunger, malnutrition and the everyday violence of 'Neoliberal Globalization' as their spokesperson Subcomandante Insurgente Marcos has called it, offer critical lessons in decolonizing food, nutrition and health. In linking up with the call to 'do nutrition differently,' I offer the lessons I have learned from the Zapatistas' non-violent resistance against neoliberalism. In order to make this offering, I have to take us back to when I first arrived to Chiapas, Mexico at the Universidad de la Tierra-CIDECI, a university dedicated to promoting education rooted in indigenous epistemology and highly respected by the Zapatista communities.

But, before I continue with that story I should offer a brief background of the Zapatistas for readers who may not be familiar with their struggle. Who are they? And, why did they declare war against the Mexican government on January 1, 1994? This is a narrative that has been told many times by many writers, journalist, scholars, philosophers and academics. Perhaps it is best if you read it from the words of a Mexican journalist who lived seven years with the Zapatistas and, as

---

1   I first came across this quote in an article by Gloria Muñoz Ramírez.

SCI Marcos said, wrote "the most complete public history" of their movement, *The Fire & The Word: A History of the Zapatista Movement* by Gloria Muñoz Ramírez:

> On November 17, 1983, a small group of indigenous people and mestizos set up camp in the Lacandón Jungle [Chiapas, Mexico]. Under cover of a black flag with a five-pointed red star, they formally founded the Zapatista Army of National Liberation (EZLN). And they began an unlikely adventure.
>
> Ten years later, on January 1, 1994, thousands of armed indigenous people took over seven municipal seats and declared war on the Mexican government. Their demands: employment, food, housing, health, education, independence, justice liberty, democracy, peace, culture and the right to information.

The EZLN is a political-military organization composed of Chols, Zoques, Tojolabals, Tzotzils, Mams, and Tzeltals among many Mayan communities in Chiapas. The indigenous women who form part of the Bases de Apoyo Zapatistas organization were among the strongest proponents of declaring war against the Mexican government. As stated in the EZLN's First Declaration of the Lacandón Jungle, the war was a "last resort, but just" against poverty, exploitation, racism and the deaths of curable di-eases that plagued their children (Ramírez 2008). Since the 1994 uprising (the same day the North American Freed Trade Agreement was implemented), the Zapatista communiqués state that each group is under constant federal, state and local military and police repression by the Mexican government for recuperating ancestral territories—rich in biodiversity, minerals, pure water and medicinal plants—therefore making less land available that is for sale to private transnational corporations and agri-businesses like Monsanto, Cargill and Pfizer. The Zapatista women continue to be at the forefront of radical non-violent resistance against the Empire of Money with their children strapped across their backs, defending and collectively working the land of their ancestors cultivating corn, beans, chiles, coffee and traditional medicinal plants. The Zapatista Women's Revolutionary Law, co-established by the late Comandanta Ramona is also as set of principles that has revolutionized women's participation as insurgents and commanders in the EZLN and as community-appointed representatives throughout all spaces of autonomous governance. Zapatista women also participate within other political-cultural spaces as promoters of health and education and participants in food co-operatives.

After my first bowl of delicious black beans and café de la olla in 'rebel territory,' at Universidad de la Tierra-CIDECI, our caravan was briefed by the local human rights defenders on the challenges faced by indigenous Zapatista communities in rural Mexico as they confront the structural violence inherent in neoliberal politics and the modern conventional corporate food system. The speakers contextualized the structural violence of Neoliberal Globalization as part of Post 9/11 "War on Drugs and Terror" national security doctrines that are used to criminalize social movements. Below I will explain how this plays out in the form of low-intensity warfare against Zapatista autonomy, in which the defense of ancestral land, water, seeds, and autonomous food production are at the center of struggle. Moreover, I will try to convey how the Zapatistas' struggle over knowledge production in the realm of health and nutrition fits into the defense of ancestral lands and ecologies. In order to witness and document how the Zapatista autonomous communities radically and non-violently resist the Mexican government's counter-insurgency tactics and low-intensity warfare, the caravan organizers divided us into four observation brigades.[2] Each brigade was sent to four of the five *Caracoles*[3] named *La Garrucha, La Realidad, Roberto Barrios* and lastly, *Morelia*, the *Caracol* to which I was sent.

There was an original 'report-back' from our brigade (first disseminated by Regeneración Radio) that documents our collective observations of the communities we visited at that particular space/time moment—four years ago to the writing of this chapter. While I did return several months later to Chiapas (San Cristobal de las Casas and Caracol Oventic) with a delegation of students from the west coast of the U.S. to participate in the First World Festival of Dignified Rage, I have not since returned to the same communities that I visited that first summer. Therefore, I cannot say what has happened since in these specific communities. What I must say is that since my last visit to Zapatista territory in 2009 the denuncias/denouncements from the JBG's published on the Enlace Zapatista website document ongoing repression against their autonomy by all three levels of the Bad Government. Audio recordings, images and documents are available in the archives at www.regeneracionradio.org that can help to provide a broader look at the communities we visited. It is also worth mentioning that I offer my collectively inspired personal account of Zapatista autonomy as a text in constant motion steadily advancing or as they say, *lento pero avanzando y caminando preguntando*. Before I proceed with my observations in Zapatista territory, I would like to contextualize my response to the idea of doing nutrition differently. It may already seem obvious that I am taking a very different approach to nutrition by offering lessons from the Zapatista movement. With our hearts and minds open, however, we can see how nutrition and health are part of greater movements for peace, justice, and dignity in Zapatista territory.

****

I had the honor and privilege of witnessing Zapatista autonomy intimately during the summer of 2008 while participating in the National and International

---

2   A group of human rights observers from Mexican, U.S., European and South American civil society.

3   A center where the Zapatista exercise their autonomous governance, justice, health, national and international gatherings, and anniversary celebrations. Each of the five Caracoles represent an autonomous regional zone in Chiapas.

Caravan in Observation and Solidarity with the Zapatista Communities in Chiapas, Mexico. This caravan was an initiative proposed to the Zapatista Good Government Councils by autonomous grassroots collectives from Europe that gathered in Athens earlier that year after an increase in military repression against the Zapatistas heightened international concerns about the situation. During the initial phase of the caravan, our brigade engaged in a series of critical dialogues that led to the formation of working groups or *comisión*. These comisiones served as an exercise in collectivity through taking on different responsibilities such as cooking, cleaning, communication and documentation that were central to the caravan. Ultimately the comisiones were critical elements in organizing the caravan for two major reasons: 1) they helped to minimize the burden of responsibility that the Zapatistas were required to take on to host us and, 2) they provided praxis-based learning about autonomous organizing, solidarity and movement building on an international grassroots level. For example, one major lesson we taught each other was through the *comisión de comida* (food). This *comisión* was responsible for gathering food from the *mercados* (markets) and preparing it for the brigade. Individuals rotated participation in the different comiciones, and when it came my turn to participate in the *comisión* de comida, I was responsible for preparing the last meal before we were sent off to visit the individual communities (Bases de Apoyo del EZLN). Here I faced the challenge of feeding a hungry brigade with firewood, a couple gallons of water, potatoes, and I think an onion or two. Needless to say some of the brigadistas from Barcelona, Madrid and Rome were concerned with the lack of meat in the meals being prepared. I took their concerns as an opportunity to check our collective privilege as people coming from urban communities where meat-based diets of greater proportion, refrigeration, and gas/electric stoves are the norm. I shared with them that as outsiders in Zapatista territory we should humble our appetites.

I should mention that this humble meal was not our only option. We did have the alternative to eat at the Café Zapatista in the Caracol where one can sit and enjoy a quesadilla, oatmeal, fruit, *frijoles negros*, tortillas, café and other snacks. There is an exchange of pesos for these *comidas*, which are collectivized by the *Junta de Buen Gobierno* for autonomous project building. However, we wondered about the implications of eating at the café for our community building. During an earlier assembly that has been called to create the *comiciones* we had gone around the circle to ask ourselves how do we want to build our community—as *brigadistas*? Some people said it would happen naturally. Other folks said that we had already started to build community the previous night during dinner—implying that that act of securing, preparing and eating foods together was essential to creating community. Someone else offered that once we had shared our own stories and experiences of resisting, then would get to know each other better. Ultimately, we came to understand that while supporting the Café Zapatista (monetarily) may still be understood as an important act of direct solidarity and mutual aid with the Zapatistas, it was likewise important and valuable to utilize the communal space created by the Zapatistas for international brigadistas such as ourselves. Utilizing the communal space—and not the least for cooking and eating—was an incredibly vital element in building the kind of international grassroots community of solidarity that we were seeking to create.

All of this—community building around food—sounds good in theory, right? Yet, putting this into praxis implied complication and hardship. Most significantly, it implied having to go out to a local market where the presence of anti-Zapatista populations, who were suspicious of our presence, made it very uncomfortable to gather foods and other necessities for our brigade. I remember encountering this discomfort when we left Caracol Morelia to gather some essentials from the mercado in the municipality of Altamirano where the presence of PRIistas[4] is strong. I remember being approached by an individual asking me, "van allá?/are you all going there?" To which I replied, "pa'onde?/where?" even though I knew he was referring to the Caracol. That exchange quickly ended as I kept walking towards our meeting point where the compas (the Zapatista guardians guiding us) arranged to pick us up to return to the Caracol. The air felt very tense and all eyes were on us. When I saw the compas arrive in the truck at our meeting point I felt a great sense of relief. Actually, it felt more like a sense of rescue—like we were being rescued from an intense confrontation that was created simply by our presence. This discomfort seems trivial compared to the common experiences of more extreme violence that the Zapatistas endure. Altamirano, in contrast to San Cristobal de las Casas where extranjeros/foreigners and Zapatista supporters are more present, certainly gave us a taste of the bi-products of counter-insurgency tactics used in low intensity warfare by the Bad Government.

****

Back at the Caracol Morelia, the Junta de Buen Gobierno (JBG) held a meeting with us to respond to some of the questions our brigade had submitted to them, and to share additional information related to the current political climate they were facing. Very quickly, the great challenges the Zapatistas face for autonomy in land, water, food and culture became apparent. For example the JBG shared with us how "military personnel have been seen on [Zapatista] territory spreading marijuana seeds" so that later "the Bad Government can accuse [them], the Zapatistas, of planting and trafficking marijuana."[5] They went on to explain how such evidence planting enables the same logic of false pretext (a 'war on drugs')[6] that legitimized the recent military incursion of *Caracol La Garrucha*, which was ordered by the governor of Chiapas, Juan Sabines. Sabines sent in over 200 agents from the federal Army, the Attorney General's Office and state and municipal police on

---

4 People who affiliate with local politicians from the PRI (Partido Revolucionario Institucional), a long-time ruling political party in many states in Mexico.

5 Speech by the JBG of Caracol Morelia.

6 The national security doctrine of neoliberal agendas like Plan Mexico (Merida Initiative) aka "U.S./ Mexico/ Central American government's War on Drugs".

June 4, 2008, just a few weeks prior to the arrival of our caravan. Meanwhile, the utilization of the 'war on drugs' as a pretext for counter-insurgency is all the more ironic since the consumption of illicit drugs and alcohol are strictly prohibited throughout all Zapatista territory.

Moving beyond the discussion of false pretext, the central theme expressed in our initial meeting with the JBG was the construction of autonomy, for autonomy and dignity (one in the same) is the objective and desire of Indigenous resistance. Indeed, despite the counter-insurgency tactics and low-intensity warfare aimed at dismantling Zapatista autonomy, the Zapatistas' radical non-violence has proven to be resilient and critical in building and defending numerous autonomous projects that include: women-led trade cooperatives, health clinics, justice systems, governing councils, education, traditional agroecological farms and cultural-ecological revitalization. Importantly, such multiple autonomous projects are designed to be interdependent, meaning that one can not exist with out the other and therefore are tied together like a braid—*una trenza*—that form an central part of the reality of the Zapatista Movement. For this reason, it becomes crucial to understand nutrition here as the cosmological relationship between healthy people, lands, seeds, water, and crops.

****

Our initial stay and meetings in Caracol Morelia also opened our eyes to the ways in which indigenous people in Chiapas have been gravely ignored by the Mexican government. Prior to the uprising, many indigenous communities were denied medical attention in hospitals and often died very young of curable diseases and hunger. To remedy this problem, the communities began to name health promoters or *promotores de salud,* and coordinated gatherings with the elders in order to enable a 'remembering' of indigenous knowledges of healing, and recollection of the medicinal properties of the local ecology. Various working groups were created to promote this 'different' way of doing both health and nutrition (for the two are very intertwined) among the Zapatista communities. While Caracol Oventic is known as one of the most advanced Zapatista autonomous health centers, I first had the honor and privilege of visiting the autonomous health clinic of the Zapatista municipality 17 de Noviembre. Our brigade was given a tour of the clinic by a *promotor de salud.* The clinic was fully equipped with a consultation room, a lab and a farmacia (pharmacy). In the center of the clinic, the Zapatista health promoter pointed out the medicinal properties found in the garden of plants, flowers, *chiles* and herbs. "*La salud autonoma es para tod@s/* Autonomous health is for everyone", he told us. Even anti-Zapatista residents in the region go to Zapatista clinics like this one to heal, and they are not denied services (another form of radical non-violent resistance). This particular clinic has established a working relationship with a somewhat nearby independent modern hospital that provides assistance to the *promotores de salud* in medically complicated or extreme conditions of particular patients.

Through this visit, we came to recognize that the Zapatista approach to medicine, their autonomous health system, is advanced and radically innovative—it elaborately involves a combination of both natural and chemical medicines to heal illnesses and dis-eases within their communities. Theirs is a practical approach that comes from the reality that after 500 years of colonial exploitation of indigenous communities, 'natural' or traditional medicine alone may not always work. While Zapatistas always start with natural medicine/methods of healing (which requires more time and often a longer healing process) they will (if absolutely necessary) resort to chemical and modern methods to control the spread or worsening of the disease and in extreme cases to prevent death. What the *promotores de salud* ultimately shared is that they are not in opposition to modern medicine; rather, they are in opposition to the capitalist nature of modern health, medicine, and nutrition. Not all Zapatista communities that we visited had the means to purchase the medicamentos químicos—chemical medications. This difference in the stages of Zapatista health also speaks to the different stages of Zapatista autonomy. Zapatista health takes a preventative approach to healing by promoting healthy eating, physical activity and soberness, not to mention a particular emphasis on the protection of reproductive health and a radical approach to promoting sexual health as well.

I share the above anecdote on health to express an important ideological approach to health and nutrition that is rooted in indigenous philosophy-principles of self-determination. Zapatistas specifically organize against exploitation by pharmaceutical companies like Johnson & Johnson and Pfizer, and agri-seed-life-science corporations like Monsanto, which have been working across the globe to patent the healing qualities of the world's rich biodiversity such as those found in Zapatista jungles and throughout their ancestral territory.[7] Zapatistas recognize their territory, their recuperated land, as the life source of their communities' autonomy and health—medicinally and nutritionally—and it must be defended by a particular set of politics and ethics (see Auirre Rojas 2008, Ch. 4, p. 171). Each community exercises its autonomous power through naming its own representatives who voluntarily take *cargos*—charge—and become *responsables*—responsible—in ensuring the community needs are being met. These responsables can propose alterNative projects, but the community will have the ultimate decision making power.

The most important *cargo* of the responsables is to make sure the land is being worked, for this is the mainstay of Zapatista autonomy. Or as they say, "la tierra es para quien la trabaje/the land is for those who work it." Despite the fact that

---

7   In saying this I feel compelled to mention that in using the term 'territory' to describe the ancestral land of the Zapatista communities I do not mean to imply a hierarchy or a notion of ownership or dominion of the land, plants and animals. In fact, I want to place an emphasis on the defense of ancestral territories against neoliberal capitalist exploitation, and I do not want to imply that there are "natural resources" or commodities in these territories that need to be under the regulation of Zapatista authorities.

community members must continually, and non-violently, avoid confrontations with anti-zapatista supporters and paramilitaries who try to stop them from working their fields, maiz, frijoles, coffee, and cattle that provide health, nutrition and small amounts of funding to build autonomous projects. While growing maiz and frijoles provide the staple foods for the communities, coffee is collectively grown and sent to Café Tatawelo—Tojolabal for "entrando a trabajar la tierra"/beginning to work the land. We had the honor and privilege to visit the distribution facility of the coffee cooperative created by six Zapatista autonomous communities: 17 de Noviembre, Che Guevarra, Olga Isabel, 1 de Enero, San Juan Can Cuk, and another one near Palenque. Each of these autonomous municipalities collectively make up around three hundred and fifty members who are certified organic coffee growers by Certimex—an independent organization from Oaxaca that guarantees coffee is grown ecologically and chemical free. The 100% Arabica coffee grown by the Zapatista communities in this region is sent to Café Tatawelo in Altamirano where it is, cleaned, roasted and ground. It's not all roasted and ground at once, however. It's done in small batches, by Zapatista mujeres who run the café— where we enjoyed smooth shots of espresso made of freshly roasted and ground café Tatawelo. Café Tatawelo sells tons of coffee per year and 75 percent of funds raised go directly to the communities and 25 percent goes to the distribution center. The funds that go back to the communities are used to strengthen Zapatista autonomous projects—especially those centered on health.

Another important aspect of Zapatista autonomous health and nutrition, we learned, is connected to cattle ranching or *ganaderia*. This is, without question, ironic—that is, the colonial legacy of cattle ranching used by the Zapatistas to produce funds through sale of cattle to build their autonomy. A Zapatista compañera of the autonomous government explained in a 2013 communiqué how the imposition of cattle ranching by the Spaniards further divided the labor between men and women. The women's domestic work became even more devalued with the rise of a bourgeoisie class of men dedicated to buying and selling cattle and their biproducts. This gender-labor hierarchy was toppled with the creation of The Zapatista Revolutionary Law For Women authored by the late Comandanta Ramona and as the cattle ranches were taken over by the Zapatistas on the day of the uprising. As told by one of the *compas* many of the *finqueros* or the landowners fled in fear at the sound of gunfire, bomb shells exploding and tanks rolling on the day of the 1994 uprising. They left behind their cattle and ranches. It makes sense for the Zapatistas to continue to maintain and sell these animals to fund autonomous health. This is how *medicamentos químicos* (chemical medicines) are bought and collective projects built. I must stress again that the main staple of these communities are *maiz* and *frijoles* grown collectively and agroecologically on their recuperated lands. Thus, the cattle do not form a significant part of Zapatista foodways, but rather of their decolonial project building. Today, as SCI Marcos writes,[8] "(…) the earth that was used to fatten the cattle of ranchers and landlords is now used to produce the maize, beans, and the vegetables that brighten our tables."

Not all Zapatista autonomous communities are advancing at the same rates in terms of strengthening the health, curing alcoholism, ending domestic violence, providing education, establishing governance and implementing the Zapatista Revolutionary Law For Women (Ramirez 2008). Nevertheless their approach to healing has radically increased infant mortality rates and life expectancy superior to those of the "great cities"[9] in addition to the overall health of their communities (people, land, air, water, plants, and animals) since the uprising. This approach is also, as I have been steadily insisting, tightly interwoven with the defense of the land and the autonomous production of food—maize, beans, squash, chiles, coffee and wide range of fruits. Which brings me now to another chapter in my story of the Summer of 2008—that is, our actual arrival into Zapatista autonomous communities. The brigade visited four that summer: *Nueva Revolucion, 8 de Marzo, 10 de Abril*, and *Francisco Villa*, all of which pertain to the Autonomous Municipality *17 de Noviembre* of the *Caracol Morelia*.

\*\*\*\*

Inside the Tzeltal community, *Nueva Revolucion*, we prepared and shared our food in an outdoor kitchen space fully equipped with pots, pans, cooking utensils and fire pits and wood. On our first night in the community, after the meal, an elder prepared café de la olla (coffee brewed in a large pot that is not filtered but boiled to point where the grounds fall to the bottom of the pot which allows you to ladle off the top) and shared this café with us along with stories of his experience of the uprising in 1994. The next morning we gathered in assembly with the Base de Apoyo Zapatista where we learn about the Bad Government's (*mal gobierno*) plans to build a dam that would flood the community and surrounding communities. Zapatistas here say that various government agencies have offered them money to abandon their ancestral lands, which are rich in biodiversity and are thus attractive to foreign investors; but they are not willing to give up their rights to land, nor are they willing to let the vegetation and wildlife of their territory be destroyed, and their deep linkages between land and health severed. The dam project, we are told, is part of the larger Hydroelectric Infrastructure Plan[10] in the region that is coupled with a development plan known as Plan Puebla Panama (PPP).

\*\*\*\*

8   See EZLN communiqué of December 30, 2012: http://enlacezapatista.ezln.org.mx/2013/01/02/ezln-announces-the-following-steps-communique-of-december-30-2012/.

9   See EZLN communiqué of January 25, 2013: http://enlacezapatista.ezln.org.mx/2013/01/25/them-and-us-iii-the-overseers/.

10   Under Mexico's Federal Electricity Commission (CFE).

In the Tojolabal community *8 de Marzo* we were given a pot and some firewood. Some of the Zapatista youth were sitting next to us laughing at our stubborn and pitiful attempt to light a fire so that we could prepare our meal. Finally we were able to get a steady fire going which allowed us to cook the last of the rice and calabazas that we picked up in Altamirano. The Zapatista kitchen spaces were the common meeting areas we utilized throughout our stay, and it was here that the Zapatistas made sure we could nourish our bodies. After we ate, the Zapatista authorities summoned the community to a meeting with the sound of a *caracol-a conch*. In addition to discussing their progress in creating autonomous education, in this meeting the community also talked of their concerns surrounding plans for eco-tourism, which plays a significant role in the PPP. After the meeting, we are taken to a sacred site, La Laguna, where the Zapatistas share the story of the creation of this lagoon, following a young girl's disappearance. Their story says that once a week a mermaid appears in the center of the lagoon with a beam of light that shoots up into the sky before the sun rises. This sacred lagoon along with an un-excavated ancient temple face the invasion of eco-tourist developers, which imply the displacement of the Zapatista communities from their ancestral territory and the conversion of this sacred site into a tourist attraction.

Later, the community continued to address their concerns over territorial control. Many concurred that neither military nor police officials directly enter the community but rather their concern is over individuals "undercover as civilians, who we suspect belong to government intelligence groups, since they arrive with the intention of knowing how we collectively organize our communal lands."[11] The community also addressed concerns over newly forming paramilitary groups. These groups are composed of individuals who "still have faith in political parties" and are being advanced and armed by the "Bad Government-[the Empire of Money]." These paramilitary groups are assisted by political elite groups that act in accordance to the security cooperation agreement between the U.S. and Mexican governments, Plan Mexico aka the Merida Initiative (Bricker 2008). We were told that the Bad Government uses the paramilitary as scapegoats in the killings, massacres, and inter-communal violence in Chiapas, while arguing that the overwhelming presence of Mexican Federal Army troops and the Federal Preventive Police is serving as a necessary "stabilizing force" for the region (Stahler-Sholk 2006). Such paramilitary forces have played similar roles in some of the other communities we visited, like the the Tzeltal and Tojolabal community "10 de Abril," which was formed in 1995 by internally displaced Indigenous communities. These communities joined the Zapatista organization that same year and have since been dispossessed of their land on three occasions; on all of these occasions, women, men and children non-violently resisted and returned to occupy their autonomous territory.

****

---

11   Quote from Zapatista community representative.

In the Tojolabal community *Francisco Villa*, our brigade experienced, again and again, what can only be described as Zapatista hospitality. When we arrived to this community the women ask us to give them our food so that we can attend a community assembly. During the community assembly some of the Zapatista mujeres—females—prepared our pasta with onions, tomato and carrots and add a bowl of their own-grown delicious black beans. All of this along with freshly brewed café de la olla we found, later, waiting for us in their kitchen but first, let me remember what we shared during that community assembly.

During the gathering the community authorities described how they decided to reclaim and occupy ancestral lands that once functioned as a landing field for Mexican federal military helicopters. Here the Zapatistas face constant harassment by helicopters flying over the area. In addition to these fear tactics introduced by the Bad Government, members of this Zapatista community indicated that they are aware of the creation of paramilitary groups in a nearby community known as "Galilee." The political parties that are behind this strategy, we were told, are the PRI, PRD and the Parted Verde Ecologist (the Green Ecologist Party), whom the Zapatista community representatives shared with us "incite [their] young community members to consume alcohol knowing it is prohibited in all Zapatista territory." The community representatives went on to explain how "these are tactics the Bad Government uses to discredit our movement and displace us from our ancestral lands to privatize the natural resources."

The fact that the Green Ecologist Party was involved in counterinsurgency tactics is a strategic move by local politicians to exploit the weight that Green Party has in promoting sustainability and protection of biospheres in the region. In reality, many members of the Green Ecologist Party are working against the Zapatistas in this community by actively promoting and advocating for the construction of a dam and a transnational (corporately-led) fish farm. Both of these projects would not only flood the surrounding communities of "Francisco Villa" but would destroy the flora, fauna and rich biodiversity of the region. According to this community these types of government sustainability proposals abuse the rhetoric of ecological and social sustainability and never carry out their promises of creating more jobs and improving the quality of life for indigenous communities in Chiapas. Instead, they work to benefit the interests of transnational corporations.' For the indigenous communities, the ramifications of such neoliberal projects are nothing less than cultural genocide and femicide[12] (Sanford 2008) combined with the priceless loss of access to clean water and food, and the subsequent generation of poverty, hunger and forced migration to major cities in Mexico and the U.S.[13]

\*\*\*\*

---

12 Feminicide is a political term, with a broader connotation than femicide because it holds responsible the state and judicial structures that normalize misogyny. (http://www. ghrc-usa.org/Programs/ForWomensRighttoLive/FAQs.htm).

13 Testimony from a Zapatista health promoter of the "Francisco Villa" community.

Our brigade returned to Caracol Morelia, and the next day we celebrated the birth-day of the Caracoles (five years prior). The Zapatista men prepared *caldo de res*/beef stew with freshly butchered cow done that same dame on site. In six super-extra-large metal pots over wood-burning fire pits, men stirred heavy deep brown broth with vegetables and beef. Later that evening as the communities from all over the region began to arrive, tamale and taco vendors began to set-up shop in the *caracol*. It was two days of political-cultural acts, denouncements, theatre and sports with an amazing spread of foods grown, and prepared by Zapatista *mujeres*. In between all of this celebration of rebellion, we gathered in assembly to collectively analyze what we had learned. We placed our attention on identifying the root of the kinds of violence that had been described and expressed to us. We talked a great deal about the role of governments like those of the United States, Canada, China and the European Union, which in different ways represent the political economic desires of transnational agribusinesses and corporations like Monsanto, Cargill, Tyson, DuPont, Pfizer, and Coca Cola, who in turn exert control over lands, and food and health systems. These governments and businesses legitimize the neoliberal policy structure which, operating under the banners of a 'War On Drugs and Terror,' justifies violence, genocide and widespread criminalization of social movements. In Mexico in particular, this marriage of corporate power and government policy is known as Felipe Calderon's War, a 'war on drugs' which is said to have incited over 50,000 deaths between 2006 and 2011,[14] countless missing people, and an increase in forced migration and displacement, as well as the persecution of activist journalists and the unjust incarceration of countless political prisoners criminalized for defending their ancestral lands and traditional ways of life as indigenous peoples. As I have already conveyed, the repressive force behind this structural violence is known at the local level as *el mal gobierno* (the Bad Government) when referring to the local, state and federal Mexican government agents. When talking about international repressive forces like transnational corporations, the World Bank or the International Monetary Fund, the term is the Empire of Money (El Kilombo Intergalactico 2007). Both, *el mal gobierno* and the Empire of Money, are terms coined by the Zapatistas.

It was during this final series of gatherings at Caracol Morelia that I gained a close understanding of how *el mal gobierno* acts in accordance with the Empire of Money to implement neoliberal plans like Plan Puebla Panama[15] and the U.S.-funded Plan Mexico aka "Merida Initiative" (Bricker 2008). Plan Puebla Panama and the Merida Initiative both aim to displace Zapatista Indigenous communities, and other Indigenous and non-Indigenous communities in Southern Mexico and Central America, by using low intensity warfare in order to make way for mega infrastructure development projects like hydro-electric dams and ecologically

---

14   Sicilia, Javier biographic sketch, accessed Feb 2012, at http://www.lfla.org/event-detail/709/Javier-Sicilia.

15   A neoliberal development plan funded by the InterAmerican Development Bank. For more see: CIEPAC, "El Plan Puebla Panama (PPP)," www.ciepac.org/ppp.htm.

destructive monocultural systems of food production (Shiva 1989).[16] Denouncing the implications of these neoliberal projects is one of the main weapons of radical non-violence used by the Zapatistas who see global neoliberal corporate capitalism as a threat not only to their autonomy but also to humanity and the ecological community—plants, animals, water, land and foods.

\*\*\*\*

My experiences with the Zapatista communities in the Summer of 2008 left me with a responsibility—*un cargo*—to share with others what I learned from the Zapatistas. Everywhere we went in Zapatista territory, they wanted to learn about the ways my community organizes and resists neoliberalism and capitalism. Then they would tell me, the best way you can help us is by going back to your communities and sharing with them everything you learned here. One of the lessons in Zapatismo is the idea that another world is possible—yes I know a new one already exists in the heart of the Zapatista autonomous communities. But this idea that another world is possible means that we have the power to create something different from below and to the left. This is a major lesson of Zapatismo that inspires my work as a traditional foodways chef and activist scholar developing an autonomous grassroots decolonial healing project. Through organizing community/university workshops, blogging, family dinners, workplace and everyday conversations I engage in decolonial movement building where food is my weapon of choice.

From the dialogues, the physical closeness with their autonomous health projects, the sensory experience of their lands, and the sharing of food within Zapatista territory, I came to viscerally acknowledge the deep importance of the Zapatista struggle for regaining control over our life-health systems and specifically for the decolonization of our global diets. Neoliberal Globalization, *el mal gobierno*, the Empire of Money and the corporate food regime use food and land grabbing as a weapon of structural violence (Barndt 2008; Esteva and Prakash 1998; Farmer 1996; Shiva 1989) to dismantle opposition to their hegemonic power. Indigenous communities like the Zapatistas who construct their own autonomous production of food, and their own systems of health and nutrition are criminalized under "War on Drugs" national security doctrines and therefore subject to counter-insurgency and low-intensity warfare (Bricker 2008; Gill 2004; Navarro 2008). In response to the structural violence imposed by neoliberal projects aimed at destroying the ecological and cultural dignity of indigenous communities in rural Mexico, Zapatismo is rooted in constructing radical projects of autonomy, such as alterNative health care, education and forms of governance. Furthermore,

---

16 Under the Program of Certified Ejidal Rights and Titling of Urban Plots (PROCEDE) that is commonly implemented in regions where a considerable increase in *campesino* (rural farmer) migration to the United States occurs and where *campesino* production is displaced by monocultural food production.

women's participation in the JBG's and in creating and organizing cooperatives that autonomously and collectively cultivate, distribute and prepare traditional foods offer important lessons for people at the grassroots level who defend their ancestral lands, communities and food.

****

In the radical history of the United States we can see the potential of social movements that were able to feed their communities and challenge the corporate food regime and a racist political system. For example, the Black Panther Party was feeding almost a quarter of million youth across the United States per day through the Free Breakfast for Children Program (Patel 2011). This eventually placed them as the greatest threat to U.S. "internal security" which ultimately served as FBI director J. Edgar Hoover's scare tactic to dismantle the movement through COINTELPRO. While the Free Breakfast for Children Program put men in the kitchens and in so doing attempted to confront the gender hierarchy and patriarchy within the Party, the Zapatista experience challenges us to think beyond a food system controlled by corporations. In *Survival Pending Revolution: What the Black Panthers Can Teach the U.S. Food Movement* Raj Patel provides some details as to what the "universal aspiration" of the Free Breakfast for Children Program "for a balanced diet" consisted of: "fresh fruit twice a week, and always a starch of toast or grits, protein of sausage, bacon or eggs, and a beverage of milk, juice, or hot chocolate [...]" (Holt-Jimenez 2011, 123). While we can easily fall into dialectic debates over good/bad foods in mainstream science, I'd rather see the BPP Free Breakfast for Children Program as a critical and practical lesson that teaches us how autonomous control over a localized food system go hand in hand with the self-defense and self-determination of our communities in the U.S. The Standard American Diet (SAD), which is greatly processed and meat-based, is a patriarchal-capitalist food system that dates back to colonization. You see, just as rape came with conquest, so did the idea that the brown female body we call the land and everything that inhabits her dwellings like the (feminized) animals are for the taking. Since colonization, people of color have been under colonial occupation through the foods we have been forced to produce and consume. Trapped in this colonial food matrix of power, the land and all of our relations are equally part of the same labor force that drives production and consumption of a Eurocentric Standard American Diet—a SAD diet.

With roots in a heavily meat and processed food based paradigm like the SAD, the U.S.-led corporate food regime has attempted to displace plant-based consumption within native and indigenous communities of Mesoamerica. Yet, whether plant-based or meat-based (as with some Native Alaskan and Canadian communities) health-giving Indigenous foodways that are ecologically sustainable continue to exist outside of the colonial food matrix of power. Certainly, a major lesson from the Zapatistas is one of self-determination (Alfred 1999), and how to move beyond resistance (El Kilombo Intergalactico 2007) towards decolonial autonomous movement building by remembering our traditional ways of healing and eating without dependency on the current systems of education, politics, food and health. In line with the Zapatista focus on self-determination, People of color (POC) movements in the U.S. are creating alterNative ways of doing health, food and nutrition by remembering the ways of our ancestors. Such POC movements are carrying on the ancestral guidance and answers we need to solve the problems we face today. Crucial here is any attempt to "do" health, food and nutrition differently must include the other segments of that broader braid—*la trenza*—with which health and nutrition are interwoven; so, "doing nutrition differently" also entails community self-defense and cultural-ecological revitalization, health and nutrition, autonomous food systems and governance. The Zapatistas' everyday reality exemplifies such a broadened understanding of health nutrition. There are communities of color in the U.S. that have been inspired by the Zapatistas and the Black Panther Party movement among many other to do the same. Our familias of youth, elders, students, garment workers, resilient migrant workers and street-food vendors make up the bases of support for such movements in the U.S.

Our communities may be lacking stores selling affordable fresh foods. But if you look at the geography, from below and to the left, it is a different reality. Consider the east side of Los Angeles to as far as Pomona, Califaztlan. Across these communities and in between many are considered to be "food deserts". Yet it is in this urban concrete jungle where our healthiest communities live amongst urban xinampas/floating gardens of Tenochtitlan (the problematically titled "Latino Health Paradox"[17] can explain this further). Across this geography from below you'll know and live amongst the street vendors selling home-made, locally produced food e.g. tamales, elotes, esquites, and fresh cut fruits. These vendors, like most vendors from below and across the world, are persecuted and repressed by the police and city officials who extort them with fines for not having health department permits. It is the local economy of street vending that conflicts with global capitalism. Yet, despite what those up above impose on us below, street vendors risk being fined even jailed to make a dignified living and to feed the community and their families healthy food. And if its not the vendor its the vecina/o/oa, our abuelit@s, that have been keeping the urban xinampas of our barrios thriving way before Food Not Lawns became a scene. Mirar hacia y desde abajo and you'll see, smell, taste, hear and feel it—in other words you'll probably live it—to paraphrase the words of my Hip Hop artist in rebellion compa Olmeca. Nopales, milpas, fruit trees, tomatoes, squash, and fresh herbs being grown, traded, eaten, cooked, shared, and prepared autonomously in the hood. Yet, with the banning of our peoples' his/herstories from schools, like in Arizona, and

17  The so-called "Latino Health Paradox" is scientific evidence for those who privilege Western Eurocentric research methodologies; it documents the fact that first generation migrant workers in the U.S. are healthier than some might expect. The use of the term paradox however exposes the inferiority complex of Eurocentric culture and discourse; *how dare "they" be healthier than "us"!*

the ever increasing separation of our families by ICE in conjunction with local, state, and federal agents our culture and foodways that are ancestrally healthy for our bodies, and the lands that form the basis for these foodways, will be even further marginalized and erased. Forced assimilation into the U.S. mainstream in this sense is straight up genocide. It is at this juncture—the struggle against our erasure—that the lessons from Zapatista non-violence and autonomous food and health are worth sharing.

****

In bringing this story full circle I would like to take us back to CIDECI-UNITIERRA in Chiapas, Mexico. It is December 21, 2012 "en los calendarios y geografias de abajo"[18] and the world from above is waiting for the world to end. Below and to the left we are preparing to celebrate the completion of the Mayan 13 Ba'ktun and the Zapatista Army of National Liberation's (EZLN) 18 years of public rebellion and 28 years of its clandestine birth. Early that morning over 50,000 Zapatistas marched silently into the five cities they took in 1994. This has come to be known as the largest public resurgence the EZLN has made thus far. As a Xicano, building on the bridges built by other Xicanas y Xicanos who first went down to Chiapas in 1994, *es mi deber*, my moral obligation, to propose that we ask what has/can Zapatismo teach us in the U.S.? In my humble opinion and experience, the best way to begin this dialogue is by revisiting the *Sixth Declaration de La Selva Lacandona* released by the EZLN in June 2005, which is the ideological framework that came into praxis as the Other Campaign seven years ago. In a summary by Hermann Bellinghausen and Gloria Muñoz Ramírez titled, The Next Step: The Sixth Declaration of the Selva Lacandona, they quote the declaration:

> It is an invitation to, 'indigenous peoples, workers, peasants, teachers, students, housekeepers, farmworkers, small landowners, small businesspeople, micro-entreprenuers, retirees, the handicapped, clergy-men, and clergywomen, scientists, artists, intellectuals, youths, women, elders, gays, lesbians and children to individually or collectively participate directly with the Zapatistas in a national [and international] campaign to develop a different way of doing politics, a national [and international] program of struggle from the left, and a new Constitution.

This is a declaration worth reading, collectively, and discussing. Dialogue, in particular may surround the political stance of the declaration, which affirms:

---

18  Los calendarios de abajo means the other calendars referring to the history we are making from below and to the left. Los geografias de abajo means the other geographies or communities from below and to the left in rebellion. See: http://www.jornada.unam.mx/2009/01/04/index.php?section=politica&article=007n3pol.

Never to make agreements from above and impose them below, but to agree to join forces to listen and to organize indignation; never to create movements that can then be negotiated behind the backs of those who built them, but to always take into account the opinions of their participants; never to seek giveaways, positions, personal advantage or public appointments from the structures of power or from those who aspire them, but to look beyond electoral calendars; never to attempt to solve the nation's problems from above, but to build an alternative from below to neoliberal destruction, a left alternative […]

I mentioned earlier that this entire text has been in constant motion and as I add in this very sentence, the EZLN spokesperson(s) Subcomandante Insurgente Marcos and the recently named[19] Subcomandante Insurgente Moisés (previously Teniente Coronel Moisés) have announced the EZLN will be making "changes in the rhythm and speed of [its] step, but also in its company[20] (...)

There will no longer be a national Other Campaign and a Zezta Internazional. From now on we will walk together with those we have invited and who accept us as *compas*, whether they are on the coast of Chiapas or that of New Zealand.

In this sense, our territory for our work is now clearly delimited: the planet called "Earth," located in that which is called the Solar System.
We will now be what we are in fact already: "The Sixth."

For the EZLN, to be in the Sixth does not require affiliation, membership fee, registration list, original and/or copy of an official ID, or account statement; one does not have to be judge, or jury, or defendant, or executioner. There are no flags. There are commitments and consequences to these commitments.

These words resonate and tighten the strands of our commitments that have come together like a braid—*una trenza*—we now call La Sexta where community self-defense and cultural-ecological revitalization, health and nutrition, autonomous food systems and governance are interwoven; doing nutrition differently also entails another way of doing politics where "the 'no' convokes us, the construction of the 'yes' mobilizes us, [because] here we don't want only to change the government, we want to change the world." [21]

19   See EZLN February 13, 2013 communiqué Them and Us VI. The Gaze 5. : http://www.elkilombo.org/ezln-them-and-us-vi-the-gaze-5/.

20   See EZLN January 26, 2013 Communiqué: Them and Us Part V. The Sixth. http://www.elkilombo.org/ezln-communique-them-and-us-part-v-the-sixth/.

21   See EZLN January 26, 2013 Communiqué: Them and Us Part V. The Sixth. http://www.elkilombo.org/ezln-communique-them-and-us-part-v-the-sixth/.

## Dedication

Para mi primer retoño, Huitzinatzin—our daughter-grand-mother-hummingbird who is here to teach us the old ways, to make things right again. To the next seven generations who will harvest the seeds of rebellion we plant today. To all our relations. Ometeotl.

## Acknowledgements

I'd like to give my deepest gratitude to the publishers and editors, especially Allison Hayes-Conroy who patiently and enthusiastically supported me in my humble effort to amplify the voices of the EZLN and the Bases de Apoyo Zapatistas of Chiapas, Mexico.

## References

Aguirre Rojas, Carlos Antonio. 2008. *Mandar Obedeciendo: Las lecciones politicas del neozapatismo mexicano*. Mexico: Editorial Contrahistorias.
Alfred, Taiaiake. 1999. *Peace, Power and Righteousness and Indigenous Manifesto*. Oxford: Oxford University Press.
Batalla, Guillermo Bonfil. 1996. *Mexico Profundo: Reclaiming a Civilation*. Translated by Philip A. Dennis. Austin, Texas: University of Texas Press.
Bricker, Kristin. 2008. Plan Mexico Spending Plan Released. *Narco News*. September 16, 2008. http://www.narconews.org (accessed November 17, 2010).
El Kilombo Intergalactico. 2007. *Beyond Resistance Everything: An Interview with Subcomandante Insurgente Marcos*. Durham: PaperBoat Press.
Esteva, Gustavo, and Madhu Suri Prakash. 1998. *Grassroots Post-Modernism: Remaking the Soils of Cultures*. New York: Zed Books.
EZLN. 2005. "The Sixth Declaration of the Selva Lacandona." *Enlace Zapatista*. http://enlacezapatista.ezln.org.mx (accessed November 17, 2010).
Farmer, Paul. 1996. On Suffering and Structural Violence: A View from Below. *Daedalus* 125(1) Winter: 261-283.
Mignolo, Walter D. 2002. *La Revolucion teorica del Zapatismo y Pensamiento decolonial*. Vol. XXV, in *The Zapatista's Theoretical Revolution: Its Historical, Ethical and Political Consequences Review*, Ed. Cideci Unitierra, 245-75. San Cristobal de las Casas, Chiapas: Cideci Unitierra.
Navarro, Luis Hernandez. 2008. The New Government Provocation Against Zapatismo. *La Jornada*, June 10.
Patel, Raj. 2011. Survival Pending Revolution: What the Black Panthers Can Teach the U.S. Food Movement. In *Food Movements Unite! Strategies To*

*Transform Our Food Systems*, Ed. Eric Holt-Jimenez. Oakland: Food First Books, 115-137.

Peña, Devon G. 2011. Structural Violence, Historical Trauma, and Public Health: The Environmental Justice Critique of Contemporary Risk Science and Practice. In *Communities, Neighborhoods, and Health, Social Disparities in Health and Health Care*, by L.M. Burton et al. Eds., 203-218. Seattle: Springer Science & Business Media.

*Plan Puebla Panama: War of Conquest.* Produced by Canal 6. 2005.

Ramirez, Gloria Munoz. 2008. *The Fire and the Word: A History of the Zapatista Movement.* San Francisco, CA: City Lights.

Regeneracion Radio. Audios y Cronicas de la Caravana Nacional e Internacional en el Caracol 4 de Morelia. *Regeneracion Radio.* http://www.regeneracionradio. org (accessed September 16, 2010).

Serrato, Claudia. 2010. Ecological Indigenous Foodways and the Healing of All Our Relations. *Journal of Critical Animal Studies* VIII(3): 52-60.

Shiva, Vandana. 1989. *Staying Alive: Women, Ecology and Development.* New Jersey: Zed Books.

Smith, Linda. 1999. *Decolonizing Methodologies: Research and Indigenous Studies.* New York, New York: Zed Books.

Stahler-Sholk, Richard. 2006. Autonomy and Resistance in Chiapas. In *Dispatches from Latin America: On the Frontlines Against Neoliberalism*, edited by Vijay Prashad and Teo Ballve. Cambridge: South End Press.

Zapatista Good Government Council del Caracol La Garrucha. 2008. "Mexican Military and Police Use Drug War to Attempt to Enter Zapatista Territory." *Enlace Zapatista.* June 4, 2008. http://www.enlacezapatista.ezln.org.mx (accessed November 17, 2010).

# Thematic Tabs for Chapter 12

Colonial

Discourse

Nature

Chapter 12

# Food, Community and Power from a Historical Perspective: Keys to Understanding Death by 'Lethargy' in Santa Maria del Antigua del Darien[1]

Gregorio Saldarriaga

**Editors' Note:** Saldarriaga's historical piece is relevant to at least three tabs: *colonial, discourse* and *nature*. A central lesson from this chapter is that hegemonic nutrition is not a modern phenomenon. Saldarriaga's exploration into historical narratives demonstrates historical linkages between nutrition, hegemony, colonization, and politics of repression and oppression. Ultimately, the chapter establishes the value of reading historical narratives through the lens of hegemonic nutrition – particularly with an eye to the ways in which nutritional hegemony intersects with colonial domination and political power plays. Key points are bolded to help the reader through the historical narrative.

Yesterday, as today, food is at the center of humans' everyday concerns: food tugs at the threads of our social fabric—from the physical and prosaic to the symbolic and transcendental. Starting from a specific experience that took place in a small village of Spanich conquerers in the Darién region (modern day southern Panama and northwestern Colombia) at the beginning of the XVIth century, I will try to show how food, nutrition, culture, illness, hunger, politics and production interweave in a network that must be holistically understood. In spite of the temporal and cultural distances that separate us from the events related in this text, dietetics, nutrition and food are today as closely linked to power and politics as they were in the early Modern Age. In this way, *this example of the interrelatedness of hegemony and nutrition lets us think critically about our current world.*

As different authors have pointed out, for Spaniards the foundation of cities—urban spaces—was a fundamental element that allowed them to live in society. Heir to a tradition that dates from Greeks and Romans, St. Augustine had defined for Christianity that man could only live piously and according to correct parameters *in the city*. The urban space was what gave political character to the

---

1  These pages have been enriched by the comments, suggestions and critics of esteemed colleagues and dear friends, including Pilar Gonzalbo, Solange Alberro, Luis Miguel Córdoba, Paolo Vignolo, Virgilio Becerra and Adriana Fontán.

city's inhabitants and the city was where the "perfect community" developed, in the terms of Thomas Aquinas (Pagden 1997, 32). During the Conquest's first period, Spanish cities and villages founded in the Mainland's Atlantic Coast were small towns that could disappear without leaving major physical traces. Nevertheless, it is important to take into account that the city *(polis)* was not only made by the material *(urbs)*, but also and most importantly by the human association *(civitas)* that inhabited it (Kagan 1998, 30). Conquerers created conquest and foundation communities and, with them, they integrated the Iberian Empirial order into the unknown territory through a framework of cities and villages.

Although the absence of buildings and constructions did not alter the conception of city itself—because its political character was the fundamental issue—, for some groups it was necessary to compensate for this vacuum by strengthening the community in various symbolic ways. For the purposes of this chapter, below I pay special attention to the acts in which *food* played a major role, either because food helped to join the urban community (or a part of it) or it was the cause of community fragmentation.

In 1511, in the middle of a power vacuum, the men who were settled in Santa María del Antigua del Darién asked Diego de Nicuesa, then governor of Veragua, to annex this village to his government. Nicuesa and his men were living very difficult times in the newly founded Belén; thus, the call of a town that seemed to live in better conditions was attractive. For different reasons, in Santa María, the opinion concerning the invitation made to Nicuesa started to change and turned into a generalized rejection that was socialized in a ceremony celebrated in San Sebastian Church, where...

> close to the altar, on the floor, they put a blanket or rug, and a bed pillow, and a cross above, as it is usually done on Maundy Thursday or on Good Friday when processions take place; and they solemnly took an oath, upon that cross, that they would not accept Diego de Nicuesa as their governor.

> pusieron al pie del altar una manta o tapete en tierra, e una almohada de cama, y encima una cruz, como se suele hacer el jueves de la Cena o el Viernes Santo cuando se andan las estaciones; e juraron allí solemnemente, sobre aquella cruz, que no rescibirían a Diego de Nicuesa por gobernador (Fernandez de Oviedo 1853, *Historia general y natural de las Indias*, libro 28, cap. III.).

Choosing the church to celebrate this oath of rejection and, at the same time, of union, is very meaningful, because the idea of a perfect Christian community was stressed: the social body of the city grouped in Christ's mystic body. While the presence of the human association is clear, the presence of food is less visible. Fernández de Oviedo was not present at the ceremony—it ocurred years before his arrival to American land. Nonetheless, he read the oath record drawn-up at the moment and was informed of the participants; in addition, he must have been familiar with this type of ceremony. Therefore, it is feasible that the words he chose really reflected the act. Thus, the references to the Maundy Thursday and the Last Supper can be revealing because they show that the tangible food was not the social coordinating element, but an allegoric one in which the communion of Christianity was configured, a Christianity that shared the bread and wine, Christ's flesh and blood. If emphasis is placed on Good Friday, the explanation loses strength; nevertheless, the constant of the presence of Christ's body and blood is kept.

Once Nicuesa arrived to Santa María's nearest port, the oath was put into effect through actions: first, armed men stopped him from disembarking; only after many pleas and negotiations, could he reach land, but without taking office as a governor, just like any another man. Nicuesa was lodged in Vasco Nunez de Balboa's house, and both shared food and shelter (Fernandez de Oviedo 1853, libro 28, cap. III). For Fernandez de Oviedo, this closeness was part of Núñez de Balboa's strategy to access power and get rid of Nicuesa, whom he sent to a certain death in a riddled ship incapable of enduring a trip to Santo Domingo or any of the Antilles. Although this supposition might be true, it is also possible that this understanding of events may have been an interpretative canon promoted by Pedrarias Dávila against Núñez de Balboa.

Independently from Nicuesa's tragic end and Núñez de Balboa's intentions when approaching him, *food was an effective way of establishing camaraderie and fellowship relations.* For instance, Núñez de Balboa cared about feeding the sick soldiers of his group, trying to keep a nutritional balance; he established horizontal and vertical links that were strengthened through sharing or providing food. This strategy could be partially based on the his own passage from being a stowaway, hidding in a barrel of flour, to being a captain. His rise was not due to legality (of one or another capitulation), but to his legitimacy among his group—a general consensus that made him the head of the conquering process put forward in the Mainland (in the moment of a power vacuum); he depended on his capacity to establish personal and group contacts and, therefore, he needed to get partners with whom he would share bread and food. Such strategy also had its origins in a very old popular tradition which associated living together with eating together. Using Massimo Montanari's expression, "on the table" as a methaphor of life (Montanari 1993, 94), Núñez de Balboa, in order to strengthen the basis that supported his power, tried to use food as a political tool through a more or less egalitarian system of consumption.

Soon after, in 1514, four years after being founded and three after Núñez de Balboa was its highest authority, Santa María del Antigua, the Europeans' westernmost settlement in the Indies at the time, was the scenario of great mortality among Spaniards: in less than a month, between 500 and 700 people died of "lethargy" (Fernandez de Oviedo 1853, libro 30; "Relación que da el adelantado de Andaboya..." p. 106; DIHC, tomo I, p. 53).

The first question that arises is: What is lethargy? I have not found a conclusive historiographical answer; even the epistemological conceptions of the History of the Sciences consider this question impertinent and meaningless. Other approaches

have tried to unmask the possible existing diseases behind such name and have
concluded that it could be typhus or even yellow fever (Crosby 1988, 110, 111,
116).[2] I do not think it is either one, because while there are no clear descriptions
of the disease in the chronicles of Santa María, the little they say does not seem to
refer to a contagious illness: even though many people died in a very short period
of time in that place, there are no reports about people following practices tending
to avoid the transmission of lethargy or about the disease spreading to the native
population of the region, just as nothing indicates elsewhere, when speaking
about lethargy, that it is an epidemic disease.[3] In turn, Carmen Mena (2003) has
pointed out that lethargy can be considered a set of diseases, which is, no doubt, an
explicative possibility. However, *I will pose a different idea of lethargy—an idea
centered on food and hegemonic nutrition*—based on what has been said about the
phenomenon in the Modern Age and what is registered about mortality in Santa
María del Antigua del Darién.

We cannot count on a clear description about lethargy from within Santa María
because, when announcing it, the codes of communication at the time allowed for
understanding what was happening without the need for detailed explanations.
The challenge of accessing lethargy is that I have not been able to find, in the
medical literature of the time, the characteristic features of the disease that would
enable a comprehensive means of analysis. Nevertheless, amidst such ignorance
it is necessary to get closer to what it is known. I insist: lethargy does not seem to
have had contagious or viral characteristics. There are descriptions of the disease
in other contexts: for example, at the beginning of the XVII[th] century, a woman,
after being ill during nine days, "Dawn came [...] and she was in deep lethargy,
speechless, unconscious and unable to wake up, snoring with her mouth open and
breathing heavily"[4] Also, Judge Juan Rodríguez de Mora maintained that he had
suffered from fever and lethargy in Nombre de Dios, in his transit to his destiny,
the New Kingdom of Granada (Rojas 1965, 28)[5]. Lethargy also appears as part
of a military confrontation, without mentioning its characteristics. This happened
years before the Darién's experience, in the conquering process of the Canary
Islands: one of the last strongholds of the Guanche resistance died of lethargy in

2   Some of the works that state that it is possible to identify diseases according to the
descriptions of the time are found in Cook and Lovell (publishers) *Secret Judgments of
God*; Toledo (n.d.) I thank Virgilio Becerra for sending me the latter.

3   In 1571, Venero de Leyva wrote that the aboriginal population of the New Kingdom
of Granada had been affected, during the 15 preceding years, by smallpox, fevers and
"lethargy, side pains, nits behind the ears, swelling and diarrhea with blood", quoted by
Villamarin and Villamarin (1992) p. 119, nota 7.

4   "Amaneció [...] con una profunda modorra, sin habla, sentido ny acuerdo alguno,
roncando con la boca abierta y el pecho alçado. " Mendez Nieto 1989, Libro 3, discurso 7,
p. 332.

5   Gelis 2005, provides a very similar description of lethargy, that is, a state of
conscience loss, among other synthoms that are not produced by lethargy but rather they
produce it.

their camp, just before having a confrontation with the conquerors (Crosby 1988, 116). The descriptions and moments in which lethargy appears make me think that it was not a disease itself but a symptom of many possible diseases; it was an agonic state—in the worst case—or a stage in which the person loses the capacity to express him or herself.[6]

From the narratives of Fernández de Ovideo and Las Casas, it is possible to infer that the disease was directly linked to the hunger produced by the lack of supplies in the village. This leads us to another question: Why was there a food crisis in Santa María del Antigua? Supplying the village had worked well before and, during almost two decades, Spaniards had overtaken the conquest enterprise in the Antilles without major food setbacks; besides, the intertropical zone is usually rich in mountain animals, fruits, corn and yucca. Nonetheless, at the end of 1514, the system seemed to collapse. To understand why, it is necessary to carefully observe the context: if the data that Fernández de Oviedo provided is correct, Santa María's population was of 515 men, with 1500 Indians at their service (Fernandez de Oviedo 1853, libro 30, cap. VIII, p. 232). On June 30th the same year, an army of 2000 men led by Pedrarias Dávila—who held the office of Governor of Castilla de Oro—arrived. Gonzalo Fernández de Oviedo and Pascual de Andagoya, who arrived with this group, formed different impressions of Santa María del Antigua: for the first one, it was a "fine village", with a good fish supply, while, for the latter, it was a small village, with a poor food supply, surrounded by hilly and flooded lands ("Relación que da el adelantado de Andaboya..." p. 106). Beyond these authors' impressions, we must not forget that, before Pedrarias' arrival, each Spaniard counted on three Indians at his service on average; but when the Spanish population had a five-fold increase and the number of Indians remained the same, each Spaniard had less than one Indian at his service. Thus, proportionally, each Indian's work had trebled in less than a month—in case that there had been an equitable distribution of manpower, and there is no reason to think that it was so. Nevertheless, it does give us insight into the Spaniards' dependency on the Indian population to get food in that small frontier village.

With the Conquest, the requisite work of Indians considerably multiplied, not only because of what they had to produce as a surplus for the Spaniards, but also for their own food demands, which needed to be satisfied. Iberian patterns of food consumption were mainly based on bread; *bread served as the core, the*

---

6   In 2006, in Turbo, Colombia, four men boarded a German ship as stowaways. Only two survived. One of the survivors said that, after a six-day journey, two of his companions died and he felt that it was his turn and that "I lived closeness to death approached as a *permanent somnolence that takes away the ability to think. When we are going to die, we become dumb*". *El Tiempo* newspaper, the 26th of March 2007, consulted in http://www.eltiempo.com/archivo/documento/MAM-2427856, on the 10th of July 2008. Italics added. I know that taking this experience to illustrate my comprehension on lethargy may seem anachronistic, but, somehow, Álvaro Riascos' descriptions correspond to the ones made in the XVI[th] and XVII[th] centuries.

*center and essential sustenance.* As Montanari has proved, there were two clearly differentiated models of carbohydrates and protein consumption in Europe: on the one hand, in the north, people preferred meat, while bread intake was low; on the other, in the Mediterranean, people ate a lot of bread, while little meat (Montanari 1993, 113). That is, for the Iberians meat accompanied bread and not the contrary: "Bread is the most eaten food and without it no food is brought to an end",[7] since "If bread lacks, we cannot eat other foods".[8]

As wheat was not available in the new conquered zones of the Mainland or the Antilles, Spaniards were obliged to turn to breads made of products of the soil like corn and cassava. The comprehensive analogies established by the Spaniards between wheat and corn breads, made corn the wheat of the Indies. Although wheat bread and corn bread are not similar, the bread function made them equivalent in times of need, since "the food's morphology is what guarantees the continuity of the system" (Montanari 2006, 103).[9]

Corn breadmaking demanded effort and work of female Indians, as the women soaked the corn, milled it with stones and hands, knead the dough and, finally, broiled it so the bread was hot at the moment of consumption.[10] In the framework of traditional Indian societies, this female work was exigent; moreover, the multiplication of consumers and the progressive diminution of labor force made the work even harder. In 1546, Miguel Díez de Armendáriz pointed out that corn bread was "hard to make and so costly to the poor female Indians that it seems a heavy weight to the conscience".[11]

Although the proliferation of consumers itself increased the work of female Indians, there was another factor that explained this increase even more. The new impositions of the dominant society required *greater bread consumption* than the one Indians used to have, since they consumed the corn in many other ways: still tender in the cob, in a fermented beverage (*chicha*), porridge (*mazamorra*), and mush (*masato*) as well as bread. It is difficult to estimate the percentages of consumption for each of these forms; however, it is probable that most of the consumption was made in fermented beverages and porridges, while breads were just one way of consuming corn among many other possibilities, and certainly not the most important. On the contrary, for the Spaniards, bread was the main form of sustenance and, even though they incorporated the other options to their diets, it was to a much lesser extent.

---

7    "El pan es lo que más se come y sin él ningún mantenimiento se acaba," Mendez Cristobal 1991, 117.

8    "Faltando el pan no podemos comer de los demás alimentos," Huarte de San Juan 1976, 249.

9    "La morfología del alimento es la que garantiza la continuidad del sistema" Montanari 2006, 103.

10    An excellent description of this process can be read in Cey 1994, 23.

11    "Trabajoso de hacer y tan costoso a la salud de las pobres indias, que parece gran cargo de conciencia", DIHC, tomo VIII, p. 181; Lopez Medel 1992, p.153.

According to Fernández de Enciso, the Indians from Urabá and the Spaniards that lived there drank *chicha* and it kept them fat and healthy (Fernandez de Oviedo 1853, Sumario, cap. X, p 97; Fernandez de Enciso 1974, 267). It is very likely that the conquerors were indeed used to this type of carbohydrate consumption, due to the years that they had lived with the Indians. Nevertheless, newcomers were still firmly attached to the consumption pattern that did not include carbohydrates in the form of liquids: they had to be solid—baked or broiled. I should also point out that the half-thousand dead (from lethargy) came from the newcomers that arrived with Pedrarias, not from the ones settled; that is, those who died were firmly attached to their bread-based consumption patterns. In this way, consuming more or less bread was not only a form of (carbohydrate) ingestion; *the choice to base one's diet around bread (or not) implied an interpretative canon that categorized human beings according to their food habits,* and served to place them close to or to distance them from the dominant (Spanish) culture, and even encouraged a social recognition of positive attitudes in those who ate more baked carbohydrates (Fernandez de Enciso 1974, 267).

If the Indians' work had already increased with the Conquest, the arrival of Pedrarias multiplied it significantly. Therefore, on the one side, we have a productive limit (the labor force) and, on the other, a Hispanic comprehensive limit (a need to consume solid breads, in spite of the many possibilities). Nonetheless, those two limits were only part of the answer at this point, since, according to Fernández de Oviedo, the maize crop was eaten by a locust plague, which indicates an additional productive limit. In the warm lands of the intertropical zone, there were two maize harvests a year: one that was sowed in the middle of March and gathered in July (around Saint John's day) and other that was sowed around August and gathered at the beginning of January (in Christmas). It is probable that, as Pedrarias' fleet arrived in June 1514 and the mortality was at the end of that year, the lost crop was Saint John's and not Christmas'; shortages were not experienced immediately, but with time, once the maize reserves were completely exhausted. This element reveals another food comprehensive limit of the Spaniards: having been just a few years in American lands, they depended enormously on the maize production because it was one of the elements that allowed them to keep a diet high in bread carbohydrates and, when it failed, their provision system tottered.

It was previously mentioned that there was another bread preparation that the Spaniards consumed: the cassava bread. As many other aspects of the American conquering experience, the Antilles stage was marked by the use and consumption of the yucca for Spaniards; thus, yucca and cassava bread were an unbreakable pair for Spaniards, unthinkable one without the other. Europeans observed that, if the final product was for immediate use or a delicate food, it was prepared thin. If, on the contrary, it was meant for navigation (to be stored for a long time), or it was a coarse food for the servants, it was prepared thicker. Nevertheless, in general terms, conquerors did not like this bread, because they considered it very dry and not palatable; even Juan de Cárdenas pointed out that eating it was like "eating sawdust" (Cardenas 1988, 162). Among all those who described the cassava bread, the need to soak it in broth or to swallow each bite with water to make it eatable was emphasized (Ocaña 1987, cap. I, p. 36; Monardes 1990, 293; Lopez Medel 1992, cap. 6, p. 154; Carletti 1976, 28). In these descriptions, the consideration of resistance over time ended up weighing more than flavor and taste. This bias might not come exclusively from an ethnocentric perspective, but rather from an established hierarchy according to the commercial factors of the time, which valued durability over taste.

The carbohydrate par excellence in European ships and regiments before and after the discovery of America was the biscuit, a long-lasting wheat bread made without yeast that could be eaten even over the years (Ritchie 1986, 126-128; Piqueras 1997, 21). With the existing difficulties to grow wheat in warm places, access to biscuit was expensive and difficult during the most important period in the West Indian enterprises. With this outlook, in the Antilles, the durability of the cassava bread, instead of the biscuit, made it the ideal food for supplying the troops that went in search of new territories to conquer or the ships used for Atlantic trade.[12] A breadmaking model through which war and long-distance transportation could be kept up, cassava bread served the purposes, through a comprehensive analogy (cassava bread: biscuit) sustained on features, functionality and possibilities. For this reason, in times of need, Santa María del Antigua asked the Court of Santo Domingo to send, among other products, cassava bread, but it is not mentioned among the ones produced for consumption in the Darién.

For the American aborigines of the Antilles and the warm coastal zones of the Mainland, yucca was a complement to maize growing and consumption. The ground used to grow maize had some yucca trees in between and, above all, its roots, which were consumed in different ways. Unlike maize, yucca recollection does not require such promptness: the roots can remain buried time after they are ready, without getting damaged or old. Thus, when a maize harvest was ruined, it was always possible to count on the yucca one. In the Antilles, they grew *Caribbean yucca* (*Manihotesculenta Crantz*), which is poisonous in its natural state because it contains cyanohydrin acid (Lovera 1988, 34). In order to make it harmless and nutritious, this yucca must be processed by peeling and grating it (and cooking it to make cassava) (Cey 1994 21-22; Fernández de Oviedo, *Sumario*, cap. V, p. 71). This bread could last a long time, approximately a year—according to Fernández de Oviedo—, if it did not get wet, because, in such case, it had to be eaten immediately. This storing possibility made it such that the cassava bread was served as a favorable food for those times in which other products with less durable characteristics, like maize, were not available.

When conquerors passed from the Antilles to the Mainland, they started to notice certain cultural differences and similarities among groups, and land productive

---

12   "[…] están estas tortas mucho tiempo sin corromperse, y las traen en las naos que vienen de aquellas partes, y llegan a España sin corrupción, y sirven por bizcocho a toda la gente". Monardes 1990, 293; "[El cazabe] es el ordinario bizcocho con que se navega en estas costas", Simón 1982, tomo III, p. 73.

differences. On one side, the native towns of the east side of the Magdalena River produced and ate cassava bread as in the islands. Meanwhile, the towns of the west side—where Santa María del Antigua del Darién was settled—did not have cassava bread in their dietary and productive repertoire: they did not ignore the yucca but instead counted on a non-poisonous variety called sweet or *boniata*. It is not necessary to transform this type of yucca into bread, since it is eatable with just cooking it over a wood fire, in a cooking pot or prepared as a fermented beverage (Fernandez de Enciso 1974, 269; Cey 1994, 19-20; Pinzon 1994, 297). For the Spaniards, this meant that they could not count on lasting bread that would allow them to organize expeditions or sustain people when there was a shortcoming of other products, as it happened in Santa María del Antigua.

Producing cassava bread on a domestic scale demands a great amount of effort due to the grating, pressing and broiling; And, producing it on a mass scale, with the typical bread demands of the Mediterranean culture, greatly increased the task. Therefore, the imposition of this new practice to the Darién natives was not simple (nor was it easy to the natives of Cartagena years later). Much of the trouble lied in the number of labor force (to make cassava bread) and its ratio to the consuming population. It is clear that, in the crisis, *naborías* were not supposed to start making cassava bread, but that there must have been a cassava supply that would help relieve the needs at the moment; as this bread was not a part of the productive cultural references of the Darién Indians, there was no cassava bread to sustain Santa María's inhabitants. While there was plenty of labor force to satisfy the supply of a population that was relatively reduced and accustomed to the land's supply, the absence of breadmaking was not that serious; but when the Spanish population quintupled, the results were catastrophic. Again, productive limitations and symbolic appropriations combined in such a way that *disallowed Spaniards from consuming foods that would relieve their hunger because there were not prepared in a suitable manner (of sustenance and taste) according to the European cultural patterns.*

A doubt clearly arises at this point: How could people who were at the seaside of a tropical zone die from hunger, if apparently a great variety of food—like mount animals, fish, lizards, eggs, fruits and vegetables—was available almost everywhere? However, all of these potential foodstuffs did not solve the problem, not by ignorance—it is enough to read Fernández de Oviedo's summary to see the number of edible things that the Spaniards knew and the values they assigned to them—but because Spaniards did not consider them nutritious or had not registered them in their alimentary codes so as to make them a common and recurrent food. Although there were many fruits available, they did not make up the basis of any western diet in the Modern Age because, among other aspects, some prejudices of the Galenic medical science carried weight for them; this science considered that an excessive consumption of fruits was a source of illness and a factor that caused an imbalance of the bodily humors (Grieco 1996, 133; Montanari 2008, 72-73). Besides, it was necessary to accompany these foods with bread, because it permitted their consumption. Some years later, Cieza de León pointed out that palm hearts were a beneficial food, but dangerous if eaten without bread, because the consumer would distend and die (Cieza de León 1984a, primera parte, cap. VI. p. 16). Many foods could be available, but if they were not adequately accompanied, they were not consumable; *ingesting wild products with bread was equivalent to civilize them, domesticate them, remove the hazards they embodied,* since bread was the cultural food per excellence and its company integrated the strange products to the social order (Grieco 1996, 479). One of the greatest problems in Santa María in the peak of the crisis was the absence of bread. It must be remembered that, according to Friar Bartolomé, people died from lethargy while asking for bread; it is possible to interpret this as a call for generic food, but I think that it was a desperate call to obtain the one food—bread—that linked them to their diet, and *the basis of their interpretative canon of what food should be.*

In addition, we have to take into account that, although the land was fertile, it was fertile within the terms of what warm and humid intertropical zones dictate: that is, some foods can easily be obtained, either by farming or hunting and fishing during all year long, but keeping them (from spoiling) is quite difficult due to the temperature and humidity. If there is no labor force available nor cultural elements (knowledge, understandings and interpretations of food) to produce and enable food-items to become and remain edible the qualifications of the land become negative.

Vignolo has recently stated that Santa María del Antigua del Darién was a laboratory in which different aspects—of what the conquest of the Mainland was to become later on—were experimented. In the particular case we have been examining, this notion of experimentation is useful when thinking about the restraints of the West Indian experience when conquerors passed to the Mainland, for their inability to adapt in crucial moments. After this catastrophic experience, Spaniards imposed a transformation of the Caribbean milieu and its diets, having as a basis a West Indian productive model. It had both native and Spanish elements, but it was Indian in essence, irreducible in itself to the previous experiences and born in the framework of the Spanish and western expansion.

There is still an unresolved element of the question related to the hunger associated with lethargy. Up until now, we have been focused on foods of American origin. However, not all of the Spaniards' food supplies depended on local (America grown) foods because, in part, the ships arrived loaded (although not full) with provisions and also because communications with the Antilles were relatively fluid, as to have a timely food support that would have freed the inhabitants of Santa María del Antigua from a precarious situation like the one they lived. These were possible alternatives that did not work out. Why?

Andagoya argues that most of the food that Pedrarias' fleet brought arrived rotten. This could be true, no doubt, since food conservation was a complicated task that, in the XVI[th] century, did not offer certainties. For his part, Fernández de Oviedo, who was much more belligerent, thought that governor Pedrarias Dávila and factor Juan de Tavira were the ones to blame: shortly after their arrival, they ceased giving the necessary helpings to the members of the recently disembarked army, instead storing the food, which was later reduced to ashes after the *tambo* (where it was stored) caught fire (Fernandez de Oviedo 1853, libro 30. DICH, tomo I, p. 53).[13] It is pertinent to remember at this point that, according to Andagoya, the ones who died were the newcomers that arrived with Pedrarias' fleet, to whom the previously settled did not help. Furthermore, those greenhorn newcomers were more prone to diseases and less used to regional meals than the ones who had already gone through an acclimation process (Vargas Machuca 2003, 82). In addition, the food restrictions imposed by the governor and the factor, and the later loss of the European supplies were blows for the newcomers and not for the previously settled ones.

Those previously settled did not help to solve the problems of the hungry because—if we recall back to the beginning of this chapter—they belonged to a group that was disputing the land and the territorial control, which they felt they had earned by their actions. *Therefore, hunger, lethargy and death were part of a political confrontation between opposing sides.* The governor did not try to satisfy the needs of the dying people of his group for many reasons, among others: as a group, they did not have an identity—like Vasco Núñez de Balboa's people; there were political and economic elements that were negotiated by means of food distribution and rationing, with which Pedrarias Dávila reinforced his power and authority conferred by the Crown. This may explain, in part, the delay to ask Santo Domingo for help. Although the mortality lasted a month, the hunger framework in which it was present was evident before the first deaths.

While for Balboa, power was achieved through the construction of a moral community that shared food, for Pedrarias food was a weapon for coercion that reinforced his authority, not by sharing it, but by restrictively managing it. The strength of the royal designation was superior to his army members' will or wellbeing: they had not designated him.

In part, the non-communitarian administrative use of food was caused by the high costs invested to achieve the Conquest enterprises. Therefore, it was necessary to guarantee the highest possible output, which would allow to cover debts and obtain profits. Even though this aspect was important, it was not determinant, because what was really at stake was the concept of the group. Something as consecrated by tradition as the sharing of food, and with it, collective identity and the recognition of others as equals did not fall apart only because of monetary trouble. After all, almost all the conquerors were in debt; even Balboa himself fled from Santo Domingo, among other things, for the debts he had contracted. The breakeven point was marked by the role that the Crown played and its position regarding the people who wielded local power. It was the contraposition between the two models: on the one side, the collective one that shared and elected; on the other, the imperial one that administered the power vertically (Zola 1984, 113-134).

---

13  According to Mena (2003) in Pedrarias' expedition there was a supply for 16 months.

The discriminatory, hierarchical view of the group existed solely for those who held power because, for the rest, the *idea of building a community through certain practices, among them the equitable sharing of food and certain inherent charity*—as for example, giving food the the ill and dying people without expecting anything in return—was still operative. We know that the group thought about this because, otherwise, it would have not bothered to mention it in their disputes, trials or protests against governors nor put so much emphasis on it. From the response of the Council of the Indies, which dealt with the protests, it is also possible to know that the non-equitable distribution of the food was an improper behavior that should be investigated, either for the physical danger that it represented for the subjects, or in terms of moral practices.

At the beginning of the XVI<sup>th</sup> century, the two models of governance coexisted as authority formulas, even making the transition from one to the other, although not in an unidirectional manner. While in Santa María del Antigua del Darién there was a transition from the group (community) model to the imperial one (from Balboa to Pedrarias), in Santa Marta it was the other way round, because the men rebelled against the governor and elected one themselves. As the Spanish power settled and the centralization process was clearer with time, the group model for electing their leaders lost strength without, however, doing away with the protests against abusive governors who monopolized food or did not respect the communitarian principles related to food. Nevertheless, as cities became stronger and more consolidated, and with them the town councils (their political representation and the governments' main organ), these where responsible for serving the interests of the citizens, resuming the role of advocate for the community, contrary to the authorities of the governors who did not behave as founding fathers.

Town councils have a very old Ibearian tradition. This is why it is not asserted that their power surged from the group model; however, due to the consoliadation of two different powers—the administrative imperial one and the cities'—there was not space for adventures like Balboa's or for popular election; still, popular nourishment and well-being was represented (somewhat rhetorically, indeed) by town concils.

It is possible to think, in a rather schematic way perhaps, that *Balboa and the community of his men and the way they shared food formed a* civitas *without* urbs; after all, what seemed to be strong was the connection among them and not Santa Marías's miserable huts. On the contrary, Pedrarias brought *urbs* (as a possible reality to be constructed with all the materials and the means that he carried and the legality that he embodied), but he had not built a community. That is why the dead of Pedrarias' group did not mean his defeat, because the mortality due to lethargy served him to legitimate the foundation of Panama and the relocation of the political burocracy there. In this way, Santa María del Antigua (despite Fernández de Oviedo's efforts) went from being the center of Castilla de Oro in the Conquest project to a subordinate and peripheral town, until disappearing altogether due to a lack of inhabitants. Thus, Pedrarias symbolically defeated the project that Vasco Núñez de Balboa embodied and erased its physical memory from a transforming geography. Although both represented different strategies, they were faces of the same *Imperium* (Pagden 1997, 24).

*****

What does this historical picture of food, power, and community help us to understand today? We find around this intriguing disease "lethargy," key elements for understanding the conquest of the Americas, as well as Colonial expansion to the west during the first years of the XVI[th] century, vis-à-vis a lens of food and hegemonic nutrition. First, we note the importance of struggles over cultural adaptation to new environments and new kinds of provisions, along with the modification of one's own food-based cultural referents and the limits to these modifications. In other words, colonial notions of what counts and what does not count as proper nutrition *ruled* all the way down to the body—so severely limiting colonizers intake as to contribute to widespread death. Second, we note the significance of challenges surrounding the capacity to manage a food production system that was being appropriated bit by bit. In this case, it was clear that the West Indian experience did not fully match the Mainland's; yet, it actively conditioned it, either for understanding it or for modifying it later on. So notions of what is proper agricultural and food production/processing technique matter here too. Third, in broader terms, Santa María del Antigua del Darién's case also serves to help us think about the ways in which nutrition and food have often been directly linked to the cultural notion of "order," of how things "should" be. This order allows for the appropriation of foods according to a clear and understandable system for the people inscribed in that system. Thus food and nutrition speak to politics and social order, and broader notions of hegemony and control. Again, in this specific case, it can be seen that, when a group found itself outside its usual sphere, facing strange conditions and provision limits, its traditional referents became limiting for survival in a place that, although difficult, was far from being barren. We might say that 'hegemonic nutrition' in once sense failed—quite literally—because of its inability to acknowledge and adjust for contextual differences, while on the other hand succeeding as a vicious political weapon of control. Standard, dominant, or dominating views about what constitutes adequate nutrition (i.e. bread) became debilitating in the new environmental and political reality. The 'interpretive canon' of the colonizers, which categorized people according to their food habits worked to distance the colonizers from the bounty of the region. Naturally, as it always occurs in situations of hunger, illness and death, politics and conflict of interests were present, not only for the conquered communities but also for the conquering groups that were continuously struggling with their own hierarchies and dynamics of power.

Certainly, the distance between the present time and this historical example of America's conquest is considerable and, no doubt, there are elements that cannot be extrapolated without being drawn into anachronisms that could be more distorting than explicative. However, we can identify key concepts and ideas through which this historical example speaks to the contemporary call to *do nutrition differently*. For instance, the chapter helps us to ask: how do food and notions of nutrition help us to understand moral communities, and the existing connections and tensions between them; Also, how are food supplies used as a political weapons or favors? And, what are the existing cultural limits—in terms of food and nutrition—of different social groups, even in their moments of major scarcity? Finally, and most broadly, in what ways does examining social relationships with/through food help us to reveal struggles over power in multiple arenas, the creation of social difference, the 'othering' of certain groups, and conversely, the bringing together of people, the construction of shared identity, and the practice of communitarian ethics? Such questions encourage us to not only to make connections between historical examples and contemporary dilemmas of nutrition but also to reinterpret narratives, new and old, for what they reveal about the linkages between food, community and power.

**References**

Cárdenas, Juan de. 1988. *Problemas y secretos maravillosos de las Indias*, Madrid, Alianza Editorial.
Carletti, Francesco. 1976. *Razonamientos de mi viaje alrededor del mundo (1594-1606)*, estudio preliminar, traducción y notas de Francisca Perujo, México, Instituto de Investigaciones Bibliográficas, UNAM.
Cey, Galeotto. 1994. *Viaje y descripción de las Indias, 1539-1553*, Caracas, Fundación Banco Venezolano de Crédito.
Cieza de León, Pedro de. 1984a. *La crónica del Perú*. Obras completas, tres tomos, tomo I. Madrid, CSIC, Instituto "Gonzalo Fernández de Oviedo."
—— 1984b. *La guerra de Chupas,* Obras completas, tres tomos, tomo II. Madrid, CSIC, Instituto "Gonzalo Fernández de Oviedo."
Crosby, Alfred W., *Imperialismo Ecológico. La expansión biológica de Europa, 900-1900*, Barcelona, Editorial Crítica, 1988.
—— The Columbian Exchange. Biological and Cultural Consequences of 1492, Westport, Greenwood, 1972.
"Descripción de la gobernación de Santa Marta" *en* Hermes Tovar Pinzón, *Relaciones y visitas a los Andes*, Tomo II, Bogotá, Colcultura, Instituto de Cultura Hispánica, 1994.
"Descripción de la ciudad de Tamalameque en la gobernación de Santa Marta [5 de marzo de 1579]", en Hermes Tovar Pinzón, *Relaciones y visitas a los Andes*, tomo II, 1994.
DIHC *Documentos inéditos para la historia de Colombia*, coleccionados en el Archivo General de Indias de Sevilla por el académico correspondiente Juan Friede, de orden de la Academia Colombiana de Historia, diez tomos, Bogotá, Academia Colombiana de Historia, 1955–1960.

FDHNRG *Fuentes documentales para la historia del Nuevo Reino de Granada, desde la instalación de la Real Audiencia en Santa Fe,* transcripción y edición de Juan Friede, ocho tomos, Bogotá, Banco Popular, 1975.

Fernández de Enciso, Martín. 1974. *Summa de Geografía,* Bogotá, Banco Popular, 1974 (1519).

Fernández de Oviedo, Gonzalo. 1853. *Historia General y Natural de las Indias Islas y Tierra Firme del mar Océano,* Tres tomos, Madrid, Real Academia de la Historia.

Gélis, Jaques. 2005. "El cuerpo, la iglesia y lo sagrado", en: Vigarello, Georges (Director), *Historia del cuerpo,* Madrid, Tauros, volume I, pp. 27-111.

Grieco, Allen J. 1996. "Alimentation et clases sociales à la fin du Moyen Âge et à la Renaissance" en Jean-Louis Flandrin y Massimo Montanari (directores), *Histoire de l'alimentation,* Paris, Fayard, pp. 479-490.

—— 1991. "The Social Politics of Pre-Linnaean Botanical Classification", en *I Tatti Studies: Essays in the Renaissance,* Vol. 4. pp. 131-149.

Huarte de San Juan, Juan. 1976 (1575). *Examen de ingenios para las ciencias,* edición preparada por Esteban Torre. Madrid, Editora Nacional.

Kagan, Richard L. 1998. *Imágenes urbanas del mundo hispánico, 1493-1780,* Madrid, Viso.

López Medel, Tomás. 1992. *Tratado de los tres elementos,* Madrid, Alianza.

Lovera, José Rafael, *Historia de la alimentación en Venezuela. Con textos para su estudio,* Caracas, Monte Ávila, 1988.

Mena García, Carmen. 2003. La frontera del hambre: construyendo el espacio histórico del Darién. En *Mesoamérica,* Vol. 24(45) pp. 35-64.

Méndez, Cristóbal. 1991 (1553). *Libro del ejercicio corporal y de sus provechos por el cual cada uno podrá entender qué ejercicio le sea necesario para conservar su salud,* México, Academia Nacional de Medicina.

Méndez Nieto, Juan. 1989. *Discursos medicinales compuestos por el licenciado Juan Méndez Nieto, que tratan de las maravillosas curas y sucesos que dios nuestro señor a querido obrar por sus manos, en cinquenta años que a que cura, ansi en España, como en la ysla española, y rreino de tierra firme,* [...] *En Cartagena indiana, año de 1607 y de la hedad del autor 76. [...],* Salamanca, Universidad de Salamanca, Junta de Castilla y León.

Monardes, Nicolás. 1990 (1574). *Herbolaria de Indias (Primera y segunda y tercera partes de la Historia medicinal de las cosas que se traen de nuestras Indias Occidentales),* edición preparada por Ernesto Denot y Nora Satanowsky, México, Instituto Mexicano del Seguro Social.

Montanari, Massimo. 1993. *Del hambre a la abundancia. Historia y Cultura de la alimentación en Europa,* Barcelona, Crítica.

—— 2006. *La comida como cultura,* Gijón, Trea.

—— 2008. *Il formagio con le pere. La storia in un proverbio,* Roma, Laterza.

Muir, Edward. 2001. *Fiesta y rito en la Europa moderna,* Madrid, Editorial Complutense.

Ocaña, Diego de. 1987. *A través de la América del Sur*, Edición de Arturo Álvarez, Madrid, Historia 16.

Oliva, Juan Pablo. 1680. *Pláticas domésticas y espirituales,* s.l., s.e., 1680.

Pagden, Anthony. 1997. *Señores de todo el mundo. Ideologías del imperio en España, Inglaterra y Francia (en los siglos XVI, XVII Y XVIII)*, Barcelona, Península.

Piqueras Céspedes, Ricardo. 1997. *Entre el hambre y el dorado: mito y contacto alimentario en las huestes de conquista del XVI*, Sevilla, Diputación de Sevilla.

"Relación que da el adelantado de Andaboya de las tierras y probincias que abaxo se ara mención (1514)", en Hermes Tovar Pinzón, *Relaciones y visitas a los Andes*, tomo I, pp. 103-186.

Ritchie, Carson I. A. 1986. *Comida y civilización De cómo los gustos alimenticios han modificado la Historia*, Madrid, Alianza Editorial.

Rojas, Ulises. 1965. *El cacique de Turmequé y su época*, Tunja, Imprenta departamental.

Simón, Pedro Fray. 1982. *Noticias historiales de las conquistas de Tierra Firme en las Indias Occidentales*, Recopilación, introducción y notas de Juan Friede, 6 tomos, Bogotá, Banco Popular.

Toledo. n.d. *Historia de la fiebre amarilla en Cuba.*

Vargas Machuca, Bernardo. 2003. *Milicia y descripción de las Indias*, Bogotá, CESO, Biblioteca Banco Popular.

Villamarín Juan y Judith Villamarín. 1992. "Epidemic Disease in the Sabana de Bogotá, 1536-1810." In Noble David Cook y George Lovell (eds), *Secret Judgements of God: Old World Disease in Colonial Spanish America*, Normand, Londres, University of Oklahoma Press.

Zola, Elemire. 1984. *Los arquetipos*, Caracas, Monte Ávila.

# Thematic Tabs for Chapter 13

*Body*

*Science*

Chapter 13

# The Nutricentric Consumer

## Gyorgy Scrinis

**Editors' Note**: This chapter links with two of our thematic tabs – *body* and *science* – as well as sets a foundation for the critical examination of hegemonic nutrition, as described in the introduction. The chapter offers an expansive critique of nutritionism – Scrinis's term coined for the nutrient-centered practice of nutrition – focusing on the taken-for-granted reductive focus on nutrients, and examining how such nutritionism has shaped the minded-bodies of contemporary consumers. The chapter helps to articulate a challenge to measurements of overweight and obesity such as the Body Mass Index as part of a broader paradigm that reduces bodily nutrition and wellbeing to a sequence of quantifiable biomarkers, and calls us to question more broadly what the body-food relationship has become.

## Introduction

Over the past couple of decades many aspects of food production and consumption has been scrutinised and politicized by food activists and scholars concerned with the health, environmental and economic impacts of the food system. The science and technology of food production has been at the centre of many aspects of this politicisation of the food system, including opposition to the use of agricultural chemicals, the genetic engineering of food crops, and the processing techniques, additives and products of the food manufacturing and fast-food industries.

The science of nutrition has, however, largely managed to escape critical scrutiny and political controversy over this period. Nutrition science has instead generally been positioned as a trusted ally by academics, public health authorities and food activists who are keen to expose the negative health consequences of highly processed foods, fast foods and industrially farmed foods, such as their contribution to malnutrition, diabetes, heart disease and obesity. Highly processed foods, for example, have been criticised for their high fat or calorie content, while organically grown foods have been celebrated for their higher content of certain micronutrients.

To the extent that there has been debate over nutrition science and dietary guidelines, this has generally been focused on the specific nutritional hypotheses and nutritional advice. The low-fat recommendation, for example, has been criticised since the 1970s by those promoting low-carbohydrate, high-fat diets, such as the Atkins Diet. Public health nutritionists have criticised various elements of the food industry for their exploitation or misuse of nutrition science to sell their products, and for their influence over the setting of national dietary guidelines (Nestle, 2007). Until recently, however, the reductive focus on nutrients within nutrition science and dietary guidelines has been largely been taken for granted (Scrinis, 2002, Jacobs and Steffen, 2003, Pollan, 2008). This reductive focus on nutrients is one of key features of what I call the ideology of nutritionism, which has been the dominant paradigm or ideology that has framed nutrition science research over the past century, dietary guidelines since the 1960s, and food marketing since the 1980s (Scrinis, 2012).

In this chapter I outline some of the characteristics of nutritionism, and examine two ways in which nutritionism has represented and shaped the minds and bodies of contemporary individuals. The first is the formation of nutricentric consumers who accept the nutricentric terms of the nutritionism paradigm, and are rendered susceptible to the nutritional marketing strategies of the food industry. The second is the corresponding representation and reduction of the body and bodily health to a number of discrete biomarkers within the discourses of nutrition and obesity.

## Nutritional Reductionism

Nutrition science has been characterised by a reductive focus on, and a reductive interpretation of, the role of nutrients in the understanding of the relationship between food and the body. In the first instance this means that the *nutrient level* is prioritised over the *food level* and the *dietary level* of understanding food. Since the rise of nutrition science in the early nineteenth century, nutrition scientists have considered the 'truth' of food's health effects to be found almost exclusively at the level of nutrients, with relatively little attention paid to studying foods or dietary patterns as ends in themselves. Scientists' initial, preliminary and limited insights into the role of nutrients have often been prematurely translated into nutritional certainties. At the same time, nutrition scientists have also created and perpetuated what I call the *myth of nutritional precision*, an exaggerated representation of their understanding of nutrients, food and the body.

This claim to precision has at times led to what has later been acknowledged by scientists' themselves as flawed scientific theories and dietary advice. For example, in the late 19th century—prior to the discovery of vitamins—the leading American nutrition scientist, Wilbur Atwater, promoted the idea that vegetables were an unnecessary and costly luxury for the working class, since they contained relatively few calories in relation to their price (Aronson, 1982). From the 1970s the American Heart Association led the vilification of eggs on the basis on their cholesterol content, advice that has since been exposed as having little scientific basis (Kritchevsky, 2004). At issue here is not that nutrition scientists occasionally get it 'wrong' in some black-and-white sense, but that they consistently elevate and prioritize their latest nutritional theories over and above other ways of evaluating the healthfulness of foods, such as those based on traditional and cultural knowledge, or on sensual, embodied experience.

A distinction can be drawn between the initial reductive focus on the nutrient level, and the further reductive emphasis on the role of single nutrients within the nutrient level. The isolation of single nutrients, and the attribution of causal relationships between single nutrients and particular health outcomes, ignores the interactions that occur between nutrients within foods, and within food combinations and dietary patterns (Jacobs and Tapsell, 2007).

An example of a single nutrient guideline was the low-fat campaign that dominated the nutriscape in the 1980s and 1990s. Reducing fat intake and eating low-fat foods was promoted by nutrition experts as capable of reducing the risk of heart disease, cancer and diabetes, and as the best way to lose or maintain body weight (Sims, 1998). This low-fat advice was decontextualized in the sense that it was taken out of any food or dietary context. For example, there was no distinction made within dietary guidelines between reducing the fat consumed from whole foods or eating reduced-fat processed foods.

While the low-fat campaign has largely been discredited since the late 1990s, other dietary patterns based on alternative macronutrient ratios have taken its place, and that emphasise the benefits or dangers of one or another nutrients or nutritional concepts. Mainstream nutritional orthodoxy now frames the healthiest diet as one based on 'good fats' (unsaturated fats) and 'good carbs' ('whole carbs' or low-Glycemic Index carbs) (Willett, 2005). In terms of counter-dietary movements, the low-carb, Atkins-style diet enjoyed renewed popularity in the early 2000s, based on the idea that carbs, rather than fat, is the uniquely fattening macronutrient we should be avoiding (Taubes, 2007).

## Nutri-Commodification and the Nutri-Centric Consumer

The success of consumer capitalism lies not only in producing an abundance of objects (i.e. consumer goods) for subjects (i.e. consumers), but also in producing subjects for objects—that is, manufacturing consumers who desire and have a need for these commodified products. Nutritionism has been an extremely powerful techno-scientific ideology for the food industry, precisely because it has produced both *nutri-centric subjects* and the *nutritional commodities* that these subjects demand or desire.

The nutricentric subject has nutricentric needs, in the sense that they understand their bodily health in terms of particular nutrient requirements. These perceived nutrient requirements are usually based on mainstream dietary guidelines and daily nutrient intake recommendations, but they may also be shaped by the nutritional marketing practices of food companies and nutrient supplement manufacturers, or the advice of alternative health and nutrition experts. The process of nutri-commodification involves transforming nutritional knowledge and nutrients into nutritional commodities. These commodities may take the form of nutritional knowledge and dietary advice and services, such as diet books, weight-loss programs and dietetic counselling; or nutritional products such as dietary supplements, nutritionally modified foods and 'functional' foods, such as low-fat ice cream or cholesterol-lowering margarine.

Nutritionism provides the food industry with a template to guide the generation and design of new products or for differentiating and adding value to their products. The focus on single nutrients promoted by nutrition science and dietary guidelines is most readily exploited by the food industry, given the ease with which single nutrients can be added or subtracted from foods, and the power and simplicity of marketing single nutrients. The development of new nutrient fetishes since the 1990s—such as the latest fetish for omega-3 fats or vitamin D—has fostered the public perception of a constantly changing nutriscape, and exacerbates individuals' nutritional anxieties that they're just not getting enough of these wonder nutrients. Nutricentric consumers are thereby caught on a *nutrient treadmill*, compelled to keep up with the latest nutritional advice and nutritionally-engineered products in order to maintain and enhance their health (Scrinis and Lyons, 2010).

An example of a nutritionally engineered and marketed beverage is 'Vitamin Water', which comes in a variety of colors, flavors, nutrient profiles and suggestive health claims. Adding vitamins to water is a simplistic and nutritionally reductive approach to nutrition, combined with the consumer value of convenience. Each flavor of Vitamin Water is labeled with playful and suggestive names such as 'Revive', 'Energy' and 'Defense'. Most consumers are probably aware that they're getting little more than flavoured and sweetened water fortified with a few random nutrients. But such are the nutritional anxieties of the nutricentric person these days that even a few random nutrients might be seen as providing just a little bit of 'nutrition insurance' to compensate for hurried lifestyles and inadequate diets. Attaining this level of perceived nutritional protection also requires little effort other than choosing this particular beverage rather than the vitamin-less one next to it on the grocery store shelf.

**Biomarker Reductionism and the BMI**

The reductive understanding of food in terms of nutrients is also mirrored in reductive ways of understanding the body and bodily health, particularly on the basis of a narrow set of biochemical processes and quantifiable 'biomarkers'. Since the 1960s this reductive approach to the body has taken the form of what I call *biomarker reductionism*, which is characterized by a reductive focus on, and interpretation of, particular biomarkers of diet-related diseases and bodily health. Some of the biomarkers that have dominated nutrition research and dietary guidelines since the 1960s are blood cholesterol levels, blood pressure, and blood sugar levels. The understanding and representation of these biomarkers has tended to be reductive in a number of senses: the reductive focus on small number of easily measurable biomarkers; the tendency to interpret biomarkers as causal agents; an exaggeration of their significance; and the claim to a precise understanding of their role in bodily health.

A key illustration of this reductive focus on and interpretation of biomarkers is the obsession with blood cholesterol levels that began in the 1960s. The belief that high total cholesterol levels had a causal role in heart disease gave way in the 1970s to the view that high levels of LDL cholesterol carriers increased the risk of heart disease, while high HDL cholesterol levels have been considered protective (Rothstein, 2003). This deterministic understanding of the role of blood cholesterol in heart disease incidence has been used to recommend a reduction in saturated fat intake, due to its ability to raise LDL cholesterol.

One of the consequences of the vilification of saturated fats—based exclusively on their role in raising total cholesterol and LDL cholesterol levels—has been the advice since the 1960s to substitute butter with margarine. Nutrition experts celebrated margarine entirely on the basis of the high unsaturated fat content of its constituent oils, while largely ignoring its highly processed character. They also largely ignored the chemical transformation of these unsaturated fats into novel trans-fatty acids by the process of partial hydrogenation that is used to solidify vegetable oils (Hall, 1974). This promotion of margarine high in *trans*-fats was challenged only after studies published in the early 1990s confirmed what had been suspected by a few nutrition scientists since the 1960s, that *trans*-fats have more detrimental effects on blood cholesterol levels than do saturated fats—at least as interpreted within the classical 'good' and 'bad' cholesterol paradigm (Mensink and Katan, 1990, Enig, 1978). In this post-*trans*-fats era, margarines produced with little or no *trans*-fats are once again being promoted for their beneficial properties. Yet margarine remains a highly processed and reconstituted food, and few nutrition experts are asking what the hydrogenated oils and trans-fat have been replaced with.

While biomarkers are generally used by nutrition and medical scientists to refer to internal biological processes, I use this term more broadly to refer also to 'external' bodily measures, such as the Body Mass Index (BMI). Within the dominant 'obesity epidemic' discourse, a person with above 'normal' BMI (>25) is characterized as being at increased risk of a range of chronic diseases, regardless of their dietary pattern or exercise pattern. People in the obese range (30-34) and above (>35) are considered at greater risk than those in the overweight category (25-29). Given this deterministic understanding of the relationship between BMI and health, losing weight and reducing BMI to the 'normal' weight range is promoted as an effective and indeed necessary means for directly improving health outcomes. This reductive interpretation of the BMI can be understood as a form of *BMI reductionism* or *BMI determinism*. Like other forms of nutritional and biomarker reductionism, BMI reductionism involves taking the BMI out of context, exaggerating its role and significance, and claiming a precise and deterministic relationship between this biomarker and health outcomes.

Critics of the dominant 'obesity epidemic' discourse have questioned whether there is a direct statistical association between BMI and health outcomes, as well as questioning the deterministic relationship between BMI and bodily health. They also question the value of focusing on weight loss as a means of improving the health of 'over-weight' people (Oliver, 2006, Gard and Wright, 2005, Guthman, 2011). One challenge to this dominant obesity discourse has come from the Health at Every Size (HEAS) movement, which promotes the idea that it's possible to be "fat and fit", and that focusing on adopting a healthy diet and exercise patterns is more important than a focus on weight loss *per se* (Bacon, 2010).

## Beyond Reductionism

A way of challenging nutritionism—and more generally the authority of nutrition science—is to re-emphasise other ways of engaging with food and the body, particularly those approaches that have been systematically devalued and undermined by the dominant ideology. One such approach is to place the quality of a food—rather than its nutrient profile—at the center of nutrition research, dietary guidelines and everyday food and nutrition discourses. I refer to this prioritisation of food quality as the *food quality paradigm*, and which forms a counter to the nutritionism paradigm (Scrinis, 2012). The celebration of whole foods, or of 'real foods', has been a feature of counter-food discourses and movements over the past decade, such as the Slow Food movement (Petrini, 2001, Planck, 2006) This has often sat alongside, or has been used in conjunction with, nutritional discourses. But it is important to also recognize the tensions—and perhaps contradictions—between these approaches.

   To put forward and re-value other ways of understanding food and dietary health is not to substitute one set of certainties for another. For example, while traditional cuisines—such as the so-called Mediterranean diet or the Okinawan diet—may be generally healthful, they do not necessarily offer either specific solutions to particular health concerns, nor always translate well into contemporary contexts. Nevertheless it is important to both de-center nutrition science, and to recognize that nutricentric scientific knowledge itself needs to be contextualized and interpreted within broader frameworks of understanding food and bodily health.

## References

Aronson, N. 1982. Social definitions of entitlement: food needs 1885-1920. *Media, Culture and Society*, 4, 51-61.

Bacon, L. 2010. *Healthy at Every Size: The Surprising Truth About Your Weight.* Dallas: Benballa Books.

Enig, M. 1978. Dietary fat and cancer trends—a critique. *Federation Proceedings*, 37, 2215-2219.

Gard, M. and Wright, J. 2005. *The Obesity Epidemic: Science, Morality and Ideology*. Abingdon: Routledge.

Guthman, J. 2011. *Weighing In: Obesity, Food Justice and the Limits of Capitalism.* Berkeley, CA: University of California Press.

Hall, R. H. 1974. *Food for Nought: The Decline in Nutrition.* Hagerstown, MD, Harper & Row.

Jacobs, D. and Tapsell, L. 2007. Food, not nutrients, is the fundamental unit in nutrition. *Nutrition Reviews,* 65, 439-450.

Jacobs, D. R. and Steffen, L. M. 2003. Nutrients, Foods, and Dietary Patterns as Exposures in Research: A Framework for Food Synergy. *American Journal of Clinical Nutrition,* 78, 508S-13S.

Kritchevsky, S. 2004. A review of scientific research and recommendations regarding eggs. *Journal of the American College of Nutrition,* 23, 596S-600S.

Mensink, R. and Katan, M. 1990. Effect of dietary trans fatty acids on high-density and low-density lipoprotein cholesterol levels in healthy subjects. *New England Journal of Medicine,* 323: 439-445.

Nestle, M. 2007. *Food Politics: How the Food Industry Influences Nutrition and Health.* Berkeley: University of California Press.

Oliver, E. 2006. *Fat Politics: The Real Story Behind America's Obesity Epidemic.* New York: Oxford.

Petrini, C. 2001. *Slow Food: The Case for Taste.* New York: Columbia University Press.

Planck, N. 2006. *Real Food: What to Eat and Why.* New York: Bloomsbury.

Pollan, M. 2008. *In Defense of Food: The Myth of Nutrition and the Pleasures of Eating.* New York: Allen Lane.

Rothstein, W. G. 2003. *Public Health and the Risk Factor: A History of an Uneven Medical Revolution.* Rochester, NY: University of Rochester Press.

Scrinis, G. 2002. Sorry Marge. *Meanjin,* 61, 108-116.

Scrinis, G. 2012. Nutritionism and Functional foods. In: Kaplan, D. (ed.) *The Philosophy of Food.* Berkeley: University of California Press.

Scrinis, G. and Lyons, K. 2010. Nanotechnology and the Techno-Corporate Agri-Food Paradigm. In: Lawrence, G., Lyons, K. and Wallington, T. (eds) *Food Security, Nutrition and Sustainability.* London: Earthscan.

Sims, L. S. 1998. *The Politics of Fat: Food and Nutrition Policy in America.* New York: M.E. Sharpe.

Taubes, G. 2007. *Good Calories, Bad Calories: Challenging the Conventional Wisdom on Diet, Weight Control, and Disease.* New York: Alfred A. Knopf.

Willett, W. 2005. *Eat, Drink and Be Healthy: The Harvard Medical School Guide to Healthy Eating.* New York: Free Press.

# Thematic Tabs for Chapter 14

> *Access*

> *Discourse*

> *Science*

# Chapter 14

# Should we Fix Food Deserts?: The Politics and Practice of Mapping Food Access

Jerry Shannon

**Editors' Note:** This chapter links with three themes: *access*, *discourse* and *science*. Shannon questions the geographic and social science behind research on food deserts – that is, areas where fresh, nutritious food is difficult to find – calling us to question whether current research tends to 'fix' these areas as objects of study in particular ways. The chapter explores how the definitions and analyses involved in food desert research may deviate researchers and activists from more complex scrutiny and solution-finding that could take into account mobility, diverse market types, difference and social stratification. One of the contributions the chapter offers to 'doing nutrition differently' is to insist on more specificity and sensitivity in geographic and social science analyses of nutrition's spatiality. Further, he argues that we will not solve the dilemmas of access to nutritious foods if we only focus on the kinds of conspicuous, large-scale solutions (like bringing in supermarket chains) that current analyses favor.

## Introduction

In February 2010, First Lady Michelle Obama released a $400 million policy initiative to improve the food access of low-income Americans. The Healthy Food Financing Initiative (HFFI) she outlined was targeted at "food deserts," urban and rural areas where fresh, nutritious food is difficult to find. Citing a number of diseases linked to poor food access, this initiative pledged to "work to eliminate food deserts across the country within seven years" (Anon 2010a). While at the time of this writing, Congress had yet to pass legislation funding the initiative, the HFFI demonstrates the increasing prominence of food deserts as a way to discuss what ails the current U.S. food system. The policy is a scaling up of previously existing initiatives at the urban and regional level, most specifically the Pennsylvania Fresh Food Financing Initiative, which successfully funded the creation of 88 supermarkets throughout the state (Anon 2011). Activists and policymakers in Chicago, New York, Oakland, Washington, D.C., and Detroit have also taken aim at food deserts. Initiatives in these cities have included financial incentives to lure large supermarkets back to low-income neighborhoods, efforts to stock fresh produce in corner stores, and zoning changes meant to encourage community gardening.

The diversity of these initiatives demonstrates how fixing food deserts remains a matter of some debate. Prominent national retailers, including Wal-mart, Target, and Walgreens, have all proposed revamping their store designs in low-income urban neighborhoods to feature a greater amount of fresh fruit and vegetables (D'Innocenzio 2010; S. M. Jones 2011; Byrne 2010). In 2007 Los Angeles banned the construction of fast food outlets in some neighborhoods to encourage the opening of restaurants with healthier offerings (Stephens 2007). New York City is one of a number of major U.S. cities incentivizing the relocation of supermarket chains to low-access areas (Anon 2009). On the other hand, critics of some of these efforts have questioned whether major retail chains, which historically contributed to the closure of small urban groceries, are the best solution to poor food access, citing sustainable food systems and independent businesses as another model (Ogburn 2010; Griffioen 2011; Mertens 2011). While food deserts have brought disparities in food access renewed public attention, exactly how to remedy them remains a topic of debate.

This chapter analyzes the role that research on food deserts plays in determining the course of these proposed solutions. Using Foucault's concept of governmentality as a guiding framework, I argue that most current research *fixes* food deserts as an object of study in three specific ways: by focusing on supermarkets as a proxy for healthy foods, by analyzing abstract territory rather than populated urban spaces, and by treating urban residents as largely immobile and passive elements of the food environment. In doing so, research tends to support highly visible, large scale *fixes*, minimizing the role of small markets, individual mobility, and processes of economic and racial segregation in shaping food access (Short et al. 2007). By treating neighborhood residents as largely passive and immobile, this strategy neglects already existing social and physical networks and individuals' embodied class and racial identities. By drawing on methodologies and practices in mobilities, feminist and qualitative GIS, and participatory GIS, I suggest that these missing elements might better incorporated into analyses of food access and suggest better, though necessarily less fixed, solutions.

### Food deserts: a primer

Research on food deserts stems in part from growing interest in an "ecological model" of public health, in which the effects of the physical and social environment on health-related behaviors is the primary focus (Swinburn et al. 1999; Egger and Swinburn 1997; Stokols 1995). Rather than focus on interventions such as nutritional guidelines meant to better educate individuals, the ecological model focuses on environmental factors like neighborhood design and store offerings that shape individual behavior. This shift from a direct focus on individuals to the more indirect influence of the environment is notable, recognizing that individuals' choices are often deeply shaped by their contexts. However, the assumptions behind existing measures of the "food environment" and models of individuals' interaction with it still deserve critical examination since, as this chapter argues, they may be set up to promote some solutions over others. To begin, a brief overview of this research may be helpful.

The term "food desert" is of relatively recent coinage, first coming to prominence a decade ago through initiatives by the British government to better understand the food options available in low-income neighborhoods (Wrigley 2002; Whelan et al. 2002; Clarke et al. 2004; Cummins and Macintyre 2002).[1] In 2008, the U.S. Congress officially defined food deserts as an "area in the United States with limited access to affordable and nutritious food, particularly such an area composed of predominantly lower income neighborhoods and communities" (USDA Economic Research Service 2009, p. 1). In practical terms, most studies of food deserts identify them through criteria including excessively long distances to stores, inflated food pricing, or poor food quality in combination with measures of social deprivation like poverty rates.

Beaulac, Kristjansson, and Cummins (2009) group food desert studies into two broad categories: market basket and geographic studies. Market basket studies analyze the foods available stores, highlighting disparities in price, quality, and availability across store types or neighborhoods. Geographic studies, the main focus of this chapter, and the studies most often referenced in policy discussions, map the locations of food sources in a city or region and then compute distances from neighborhoods to nearby food stores, which are often grouped by category: supermarkets, convenience store, fast food, and so on. Stores are often identified through an existing index of businesses in an area, such as Reference USA. Individual stores are mapped through a process called geocoding, in which GIS software uses existing street maps to place these stores by their address. Analysts add politically defined areas to this map, such as census tracts, zip codes, or block groups. GIS software then computes a distance from each of these areas to the already mapped stores, often from the area's center point (or centroid). These distances can be computed in two different ways: a straight line distance, which measures distance as the crow flies, or a network/Manhattan distance, which uses road maps to compute the actual distance traveled. Many studies which rely on the latter method convert distances to time measures, counting the number of supermarkets within a 10 minute drive, for example (Apparicio et al. 2007; Larsen and Gilliland 2008; Sparks et al. 2009). Store types are most often used as a

---

1 Research and popular concern for food security and food access in the U.S. has, of course, existed for a century or more, with past initiatives including urban gardening programs (Lawson 2005), community kitchens (Richards 1893; Shapiro 1986), and sustainable agriculture movements (Belasco 1993). Food First and the Community Food Security Coalition are just two groups with decades of activism for food security and food sovereignty. My interest in this chapter is in understanding activism around food deserts as a new way of framing this work.

proxy for the presence of healthy food, with supermarkets and fast food most often representing the positive and negative ends of the spectrum.

Once computed, distances to good/poor food sources are correlated with other demographic characteristics to analyze the connection between geographic access and these other factors, such as race, income, diet related health conditions. Spatial statistics can further highlight the significance of any disparities in the data, showing whether clusters of healthy or unhealthy food options in certain neighborhoods are likely random or reflect a meaningful disparity in food options available to residents. For example, Raja, Yadav, and Ma (2008) focused on the distribution of food outlets across block groups in Buffalo, New York. Relying on count data for different categories of food businesses, they calculated the concentration of store types within certain neighborhoods. Their analysis demonstrated that restaurants were distributed more or less evenly while supermarkets and groceries stores were concentrated in only 20-30% of the city's block groups. When block groups were classified by racial makeup (black, racially mixed, or predominantly white), the study found that supermarkets and grocery stores are relatively inaccessible in non-white areas, though the authors did note a high number of convenience stores and small grocers within these neighborhoods. In Chicago, Gallagher (2006) found a strong correlation between areas identified as food deserts and majority African-American neighborhoods. Zenk et al. (2005) noted a similar pattern in Detroit, though neighborhoods transitioning from majority white to majority African-American populations had better access. Other articles have focused on urban design. Larsen and Gilliland (2008) note that the relatively sprawling city of London, Ontario had more food deserts than larger but more compact Montreal, where no food deserts were found (Apparicio et al. 2007). More broadly, Beaulac et al. (2009) point out that the strongest evidence for the widespread existence of food deserts is within the U.S., with inconsistent findings in Canada and Europe, perhaps due to the different urban forms found in each region.

Research on food deserts is relatively new. Still, its popularity as a way of framing initiatives at the city, state, and national scales demonstrates its social and political power. In part, this popularity is due to a geographic precision that clearly delineates areas lacking adequate food access, which can then be targeted for intervention. The analytical lens of food deserts also provides clear, concrete metrics to design and measure improvements to food access. In the case of supermarkets and big-box retailers, interventions to fix food deserts offer not just measurable access to healthy food, but a sizeable number of jobs and promises of economic prosperity. At the same time, the specificity of this research—its ability to *fix* food insecurity as a property of specific, politically defined areas—is also problematic. It largely fails to incorporate less easily measured aspects of food consumption, such as small stores, informal food economies, and residents' lived mobilities. In doing so, this work implicitly frames supermarkets, big box stores, and other large chain retailers as the primary solutions to food insecurity (see Guthman 2011). It also downplays the diversity of food consumption sites and practices present even in low-income neighborhoods, opting instead for a model where distance and price alone shape consumers' actions. "Doing nutrition differently" thus involves a critical evaluation of how we come to know food deserts as geographical objects and an exploration of alternative forms of analysis and intervention that might better incorporate existing food consumption practices.

## "Fixing" Food Deserts

Most maps "fix" space, isolating dynamic and complex social and natural processes in both space and time (Hagerstrand 1982; Pred 1984; D. B. Massey 2005). Despite the fact that, as Mark Monmonier has written, "not only is it easy to lie with maps, it's essential" (Monmonier, 1991, p. 1), maps are often read as a single authoritative perspective on the world, showing things as they actually are. The various choices made in constructing a map—including a map's projection, method of data aggregation and classification, geographic scale, and its use of "visual variables" like color or size to represent its data—are often invisible in the final product, making it difficult for users to read a given map as one among many possible spatial representations. While some cartographic decisions are what Monmonier terms "white lies," others can powerfully shape shared perceptions of the world by lending their authority to projects of economic development and social transformation (Wood et al. 2010; Wood and Fels 1992; Crampton and Krygier 2006). By "fixing" space, maps not only represent the world, they help construct it.

Using the term "fix" for this phenomenon carries a double meaning. First, it captures a map's ability to represent space as stable and relatively unchanging. Census maps, satellite imagery, or road maps all possess a certain durability, depicting a reality that extends beyond one singular moment. While research on issues like climate change focuses on change over time, they often do so through a series of cross sectional maps. They present space and time though a series of slices rather than as one continuous stream. Maps of social phenomena also often present space itself as stable through a reliance on unchanging political borders, like national or state boundaries, which contain the distribution of a given variable. Second, I use the term "fix" to describe maps as a kind of framing mechanism, isolating certain qualities of interest in the landscape that may then be targeted for intervention. They get the target in the cross-hairs, so to speak. John Snow's famed map of the 1854 London cholera outbreak is one famed example of this kind of fix. By displaying both reported cholera cases and city water pumps, Snow's map fixed attention on one particular pump as a source of contagion for its neighborhood. In choosing fix as a key term, I am also mindful of Harvey's "spatial fix," which describes how capitalism's internal contradictions are resolved through construction projects and other forms of development that create new opportunities for capital investment (Harvey 2006, p. 431 ff.). Inasmuch as food desert maps (and others as well) are created precisely to mitigate food insecurity while also enabling business investment and expansion, I find this an apt association.

While generalizing about any rapidly growing body of research literature is a dangerous task, current research tends to "fix" food deserts in three specific ways:

1. Through its use of supermarkets and other large chain retail sites as proxies for healthy food access, research fixes these sources as the natural solution to poor food access, resulting in less attention to both already existing and alternative future food sources within the neighborhood.
2. By aggregating access statistics to already existing political boundaries (most often by census tract or block group), research fixes food access as a quality of arbitrary and abstract geographic territories rather than of the populations that inhabit them.
3. In focusing on residents' place of residence as the point at which access is measured, these studies fix individuals as largely passive and immobile, rather than as embodied, mobile subjects who actively engage with their build environment.

Exceptions to these trends exist, and have even gained some traction within academic institutions (Cummins, Curtis, Diez-Roux and Macintyre 2007; Odoms-Young, S. N. Zenk, and Mason 2009; S. Raja and P. Yadav 2008; A. Short, Guthman, and S. Raskin 2007). In popular discourse and policy circles, however, these traits remain prevalent.

In practice, these fixes support a governing rationality that prioritizes institutional legibility over local knowledge and rational, homogenous consumers over embodied, heterogeneous subjects. They can thus be understood as an aspect of what Michel Foucault has called "governmentality" (Foucault 1991), forms of governance that rely on a conceptualization of the nation as an aggregated unit, comprised of "the population" or "the economy." Governmentality is less concerned with absolute moral standards then with the optimal ordering of society, the "knowledge of things, of the objectives that can and should be attained, and the disposition of things required to attain them" (Foucault 1991, p. 96). It has as its goal the proper distribution of objects and behaviors within the population so as to insure the future health and prosperity of the state. Knowledge of how to achieve such goals is gained and dispersed through *technologies* of governance including, though certainly not limited to, mapping of the national territory. The knowledge of the population enabled through these technologies reinforces certain norms and goals for individual behavior so as to optimize the health, productivity, and prosperity of the nation as a whole. Studies of urban sustainability initiatives have focused on how under "green governmentality" or "environmentality," subjects come to conceptualize environmentalism and sustainability initiatives in ways that align their personal interests with those of the state and broader population (Luke 1995; Rutherford 2007; Luke 1999; Watts 2002; Raco and Imrie 2000; Agrawal 2005). Similarly, Jones, Pykett, and Whitehead (2010) examine current efforts to design environments which encourage healthful, productive behavior as a "libertarian paternalism" enabled through aggregate forms of knowledge produced by and for the state (see also Mitchell 2004; Thaler and Sunstein 2003). Such initiatives do not explicitly limit personal freedoms. Rather, they seek to structure environments so as to encourage healthy, environmentally sustainable, and economically profitable choices, while conflating individuals' and groups' self-interest with that of the state.

Food desert research certainly fits under this broad approach, along with other research using the ecological model. Its focus on the role of the physical and social environment may be preferable to a sole focus on individual choice through nutrition education, dietary guidelines, or similar mechanisms. Still, attempts to alter individual behavior through environmental influence may become a "blunt and insensitive" tool lacking an understanding of "personal needs and cultural sensitivities" (Jones et al., 2010, p. 491). That is, by *fixing* urban neighborhoods as static spaces, they neglect a range of less visible, diverse practices and sites of food consumption (Scott 1998; Gibson-Graham 1996; Pavlovskaya 2004) and minimize differences linked to class or ethnicity. They thus confuse the objective space of a city's neighborhoods with its residents' lived realities, favoring the development of the former while ignoring the latter. Maps of food deserts thus have an active role in constructing the built landscape and governing the subjects that populate it. The sections below highlight three ways these maps fix food deserts and suggest some possible alternative approaches.

*The First Fix: Equating Supermarkets and Healthy Food*

The first of the significant "fixes" in food desert maps involves the way they represent food consumption options. As noted above, many studies use large groceries as the only proxy for healthy food access, citing evidence that fast food and convenience stores are more likely to offer unhealthy alternatives (Lee and Lim 2009; Larsen and Gilliland 2008; Apparicio et al. 2007; Hemphill et al. 2008; Shaw 2006; Zenk, Schulz and Israel 2005). However, equating grocery stores with healthy food access can be highly problematic (Zenk, Schulz and Israel 2005). As an example, consider "When Healthy Food is Out of Reach" (2010), a study of food deserts in Washington, D.C. Created by a partnership of D.C. Hunger Solutions, an anti-hunger advocacy group, and Social Compact, which promotes investment in inner-city neighborhoods. "When Healthy Food is Out of Reach" bases its analysis on the locations of 43 major supermarkets within D.C.'s boundaries, considering all money not spent at these locations as "leakage"—money spent outside the city (Ashbrook and Roberts 2010). However, a list of SNAP (food stamp) authorized retailers for the city in 2011 includes 440 locations—10 times the number of stores considered within this study.

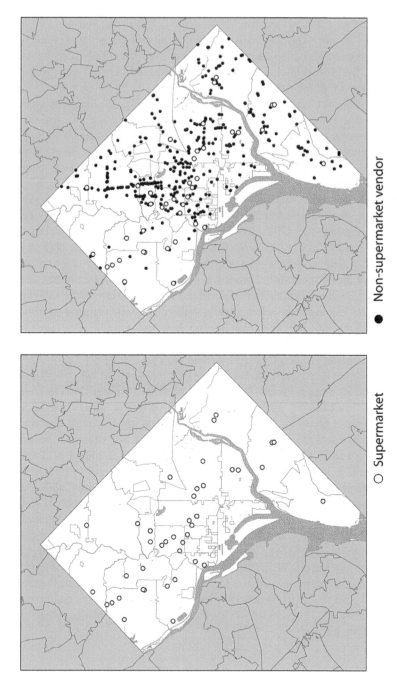

○ Supermarket      ● Non-supermarket vendor

**Figure 14.1 Of the 440 food-stamp eligible locations in Washington D.C., only about 10% are supermarkets**

*Source*: USDA.

While many of these certainly offer little beyond chips, sodas, and other processed snacks, they also include over a dozen meat and seafood markets, a handful of farmers markets and small organic retailers, and close to twenty stores catering specifically to African and Hispanic populations. As several authors have noted, these small retail sites are an often a good source of fresh food in dense urban areas (Donald 2008; Short, Guthman, et al. 2007; Raja et al. 2008), though frequent ownership turnovers and their status as independent retailers can make them easy to overlook. The smaller size of these sites makes them easier to incorporate into dense urban landscapes and thus more accessible to individuals with limited transportation options or for those without the space to store significant quantities of food. Food trucks have also garnered increasing attention from groups working on food access. The Food Truck Fiesta website (http://foodtruckfiesta.com), as of this chapters' writing, listed 49 trucks operating in the DC area. Many of these offer trendy foods targeted at white collar workers. However, Chicago's Fresh Moves bus (http://freshmoves.org) or Minneapolis' Sisters Camelot (http://sisterscamelot.org) offer free or low-cost sustainably grown produce and specifically target low-income areas.

The decision to rely on groceries as the main measure of healthy food access is understandable; a more detailed analysis of offerings in each store would make any study at the metropolitan scale extremely time consuming and labor intensive. USDA analysis of food stamp expenditures in 2008 showed that supermarkets and big box stores (Target, Wal-mart) account for a vast majority (84%) of benefit redemptions, which presumably is correlated with total patterns of food consumption in these low-income households (USDA Economic Research Service, n d, p. 62). Yet the use of supermarkets as a proxy for healthy food—fixing these as a one-size-fits-all solution—can lead to interventions that ignore other, already existing, and more accessible alternatives within low-income urban neighborhoods. Nor does it reflect how habits of food consumption relate to food availability. For example, a longitudinal study of food consumption done by researchers at the University of North Carolina found little connection between supermarket proximity and consumption of fresh fruits and vegetables (Boone-Heinonen et al. 2011). This suggests that simply making healthy foods more available in low income neighborhoods may have little effect on dietary habits and highlights the complexity of the relationship between urban residents, the foods they consume, and the sites where these foods are purchased. Decisions about where to shop and what to buy may well be influenced by availability and price, but also by the influence of cultural, gendered, and classed associations with certain foods and sites of distribution (Bourdieu 1984; Deutsch 2010; Levenstein 1988).

Part of the appeal of supermarkets and big box retailers is both their legibility and their associations with economic prosperity. They are a highly visible fix to the problem of food insecurity. Supermarkets represent economic vitality, middle-class status, and normative practices of food consumption for residents and policy makers alike. In a recent campaign to open new superstores in Chicago, Wal-mart posted a series of short videos on YouTube emphasizing two main benefits of the proposed stores: jobs and food access. One mother, small child in arm, looks directly at the camera and describes the ease of having a Walmart close by, compared to her current long drives to the suburbs. Another woman cites the benefits of lower prices, more job earnings into the community, and the influx of tax revenue from a superstore, concluding that her area has "no ... major ... stores ... *at all*" (Walmart Community 2010). In "We all are praying for this Walmart," an elderly, disabled, African-American woman recounts how in her 40 years of neighborhood residency, she saw grocery stores leave the community. Instead of "wasted," vacant land in her neighborhood, a new Wal-mart will provide nearby, affordable shopping and healthier food options. While Mayor Richard Daley initially opposed the new stores, after Wal-mart made concessions on worker wages he became a strong advocate, citing the same economic and public health benefits (Warren 2010).

The money involved in the construction and operation of these markets, along with their large food inventories and shopping convenience, make them attractive targets for politicians seeking visible signs of redevelopment in low-income communities and for residents aspiring to middle-class lifestyles. Indeed, as the Wal-mart campaign points out, these factors can be used politically to sell the value of large food markets, even when—as in Wal-mart's case—they face substantial community resistance for other reasons, such as working conditions or food sourcing. Maps of food deserts implicitly encourage this perspective, focusing strictly on distance to "fresh" food while failing to include potential alternative food sources and medium-sized/independent grocers (Griffioen 2011). In doing so, they suggest a fix which relies on current systems of large-scale, privatized food production and distribution, rather than a more deliberative and participatory discussion of how to plan for urban food. They also support efforts by supermarkets and big box stores to expand into low-income communities after the saturation of suburban markets, potentially displacing the few independent stores left in these neighborhoods (D'Innocenzio 2010; Anon 2010b; S. M. Jones 2011).

This is not to argue that the investment of Wal-mart and other retailers in these neighborhoods is wholly negative, especially given the massive disinvestment of private capital in core urban areas over the last fifty years. But it does suggest that by naturalizing supermarkets as a measure and solution to food access in urban communities, these maps normalize a governing logic in which institutional visibility and access to cheap produce take precedence over other, more complicated concerns: broader systems of foods' production and distribution, the role of agricultural subsidies on food prices, and the working conditions of those who both produce and buy food. The cultural authority of maps, the fact that they are perceived as describing the world as it is, further close off such discussion. The increasing involvement of community residents in gathering and analyzing geographic data has been noted by researchers studying participatory and public participatory GIS. Such an approach could provide a richer picture of urban food environments, as well as improving community buy in for any proposed intervention. Chomitz et al. (2010) saw significant health benefits accrue from a participatory food and health program in Cambridge, Massachusetts.[2] This might also entail the use GIS as part of a participatory decision making process on food system planning, mediating the interests of different stakeholders on the best methods of improving food access. Participatory mapping exercises might synthesize community opinion to create a visual representation of the current or ideal food system, along the lines of Hawthorne, Krygier, and Kwan's (2008) project identifying residents' preferences for new bike paths. Studies of alternative or informal economies using feminist GIS practice and qualitative methods also highlight the diverse possible trajectories of future development (Pavlovskaya 2004; Smith and Jehlička 2007). Such methods decenter supermarkets as a natural solution to food insecurity, opening discussion on a broader range of potential solutions.

*The Second Fix: Homogenizing Food Territories*

Most studies depict food deserts through the use of choropleth maps, which visualize geographic data by using color and/or texture to fill clearly bounded areal units (state/country boundaries or census geography such as census tracts). Choropleth maps are visually easy to comprehend and relatively simple to make using desktop or online GIS software. They also are easily compatible with many forms of population data which are made available in geographically aggregated form, such as census data. Like data on supermarket locations, choropleth maps are appealing for the way they clearly delineate the distribution of a given variable. Aggregating to a specific areal unit (such as a census tract or zip code) provides sharp boundaries and often clear contrasts between neighboring areas. In doing so, choropleth maps fix these areal units themselves as the object of intervention. In the case of food deserts, placing more healthy food options within a given zip code or neighborhood is a natural response to such a map. By fixing certain territory as in a state of deprivation, these maps implicitly argue for a fix tied to that space, rather than its residents.

---

2   It is worth noting that participation in this study entailed the cooperation of several governmental, educational, and health groups. It would be interesting to discover how a more grassroots approach, involving the intentional recruitment of traditionally marginalized populations at the beginning of an initiative would compare.

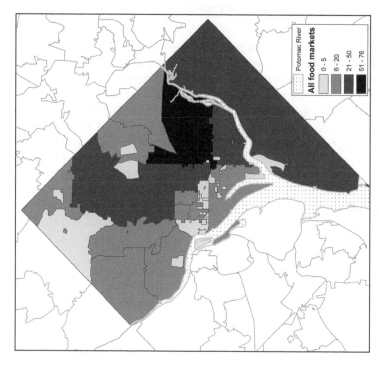

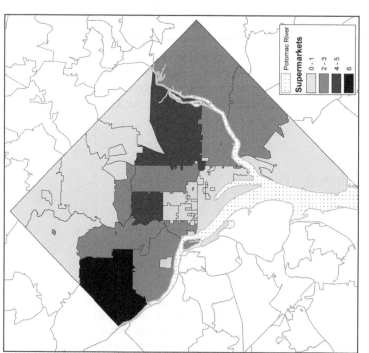

**Figure 14.2** Two choropleth maps of Washington D.C. zip codes show how supermarkets are concentrated on the west side of the city (left) even though the majority of all food stamp eligible retailers are located in the city's eastern half (right)

*Source:* USDA.

This territorial focus supports the conceptualization of the population as an aggregate, homogenous group. Jeremy Crampton terms them "part of a rationality of calculability of populations" which produces "spatially bounded conceptions of human life" (Crampton, 2004, p. 43 and 50). Choropleth maps suggest that the phenomena they represent are diffused smoothly across space while breaking sharply at what are essentially arbitrary boundaries. In doing so, they present space as comprised of discrete and stable units with clearly defined characteristics, objects whose resources and populations can thus be governed. This stability can be deceiving. Choropleth maps are subject both to the modifiable areal unit problem (MAUP) and the ecological fallacy (Walker 2010; Schuurman et al. 2006). MAUP refers to the variation in results dependent upon the scale of analysis, such as crime rates compared at the census tract and county level. Brown and Knopp (2006) describe, for example, how the scale of analysis used to map gay and lesbian populations using census data can drastically change their visibility, as households with gay men are generally more concentrated than lesbian households, making the former "easier to find and count" (p. 232). The ecological fallacy describes a situation where all members of a group are supposed to share a common characteristic, such as all residents of a low-income neighborhood being themselves low-income. In essence, choropleth maps naturalize certain scales and homogenize space, producing fixed and fixable territories.

Instead of a focus on political territories, many researchers have argued for mapping techniques that focus on populations, letting the distribution of these populations determine the scale and areal boundaries of a map. Dasymetric mapping is one often suggested alternative (Mennis 2003; Poulsen and Kennedy 2004; Eicher and Brewer 2001; Crampton 2004). By disaggregating data, dasymetric maps avoid the sharp arbitrary boundaries of choropleth maps and highlight continuity in the population across areas. A similar adjustment might be made with the current practice of using buffers to measure distance, such as the number of stores within a 1,000 meter radius of a given point. Creating a map in which distances are recorded at each map location, similar to a weather temperature map, would ensure that locations close to a boundary point—1,001 meters away, for example—are analyzed more accurately.

The territorial focus of food desert maps also supports a larger project: constructing healthful environments that indirectly influence residents' consumption choices. In this view, governing the population means not enforcing rigid rules of behavior, but creating physical spaces in which individuals will tend to make the optimal choice. Certainly, the urban "foodscape" has indirectly been the subject of governance in past years, through planning efforts which encouraged sprawl, urban renewal programs which concentrated poverty and expanded highway networks, and federal farm policies which have enabled the centralization of food's production and distribution. Yet for much of the last century, the construction and operation of urban food systems has been left largely up to private industry actors (Donofrio 2007). Choropleth maps construct this food

**Figure 14.3 Choropleth and dasymetric maps of density of SNAP clients in Minneapolis**
*Source:* USDA, US Census.

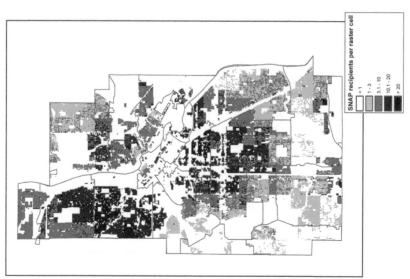

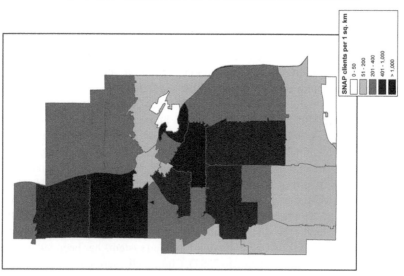

system itself as an object of governance, a territory on which resources must be distributed properly if the health of the population is to be ensured.

As a result, urban populations are homogenized by these maps, which rely on a model of human decision making which takes distance as the sole criteria in decisions about how and where to get food, assuming that nearby access to healthy food items will have a natural positive effect on diet. Improving the proximity of healthy foods to residents is thus assumed to result in positive health outcomes, a theory has not yet been validated by empirical evidence (Boone-Heinonen et al. 2011). This approach relies on an underlying framework of behavioral economics, in which individual decision making takes place within limited set of options and knowledge, while also susceptible to certain kinds of irrational behaviors, such as a preference for risk aversion (Simon 1955; Tversky and Kahneman 1974; Thaler and Sunstein 2008; R. Jones et al. 2010). While this may be preferred over the rational choice approach of classical economics, it foregrounds environmental and biological influences while giving little attention to social influences on identity and behavior (Strauss 2007). In the case of food deserts, social determinants on consumption behavior are not included: the role of culturally specific stores and foods, classed associations with some store and food types, the perceived accessibility and safety of specific stores, or the role of family and friends and shaping shopping decisions or improving personal mobility (carpooling to a suburban grocery store, for example). While a few studies have tried to take a more "relational" perspective on food environments, these have been a minority (Cummins et al. 2007). It is not as easy to model the behavior of a heterogeneous population whose food procurement habits are not so easily captured, getting a fix on the various ways people find and consume food rather than proscribing a one-size-fits all territorial solution. This approach makes discussion of social differences (in class, ethnicity, age) more difficult, normalizing certain kinds of consumption and pathologizing others. One development group, for example, describes how "in poor urban areas a culture of poverty has accumulated over time as residents embraced unhealthy lifestyles, creating habitual cycles of bad choices that are hard to reverse" (Doe, Dunlop, Sonawane and Weil 2011, p. 21). By reducing all food choices to the dichotomy of healthy/unhealthy, such reasoning stigmatizes both neighborhoods and residents and neglects other logics of food consumption linked to class, cultural, and gender identity (A. Hayes-Conroy and J. Hayes-Conroy 2008; Bourdieu 1984; McPhail et al. 2011).

Lastly, rendering urban neighborhoods as discrete, governable units obscures the historical and social relations leading to their creation. While the Wal-mart video mentioned above may frame the exodus of food retailers from core urban neighborhoods as simply a tragic event, disparities in food access in urban America have a history a century or more in the making and not divorced from Wal-mart's economic rise (Deutsch 2010; McClintock 2011; Tangires 2002; Eisenhauer 2001). By fixing the problem as a geographically specific deficiency of healthy food options, rather than a symptom of a systemic, historically rooted problem in the operation of the food system, economic policy, and urban planning, food desert maps which objectify and homogenize food territories steer attention away from these broader issues. Research linking food availability to specific planning policies may provide another approach (see Black, Carpiano, Fleming and Lauster 2011), as would historical studies that tie changes in food policy and food availability to governing strategies and economic segregation (Lawson 2005; Larsen and Gilliland 2008; Clarke et al. 2006).

*The Third Fix: Immobilizing Urban Residents*

Given their static, bounded nature, most food desert studies do little to account for the mobility of urban residents, treating places of residence as proxies for individuals' locations. An example here is research done in Chicago by Mari Gallagher and her research group (Gallagher 2006; Gallagher 2009). Gallagher's work is worthy of mention due to both her unusual attention to specific health outcomes and the subsequent studies she was commissioned to complete in Detroit, Birmingham, and Savannah, Georgia, marking her as a prominent national figure in this research field. Similar to other food desert studies, Gallagher's Chicago study mapped the density and distance to large and small grocers and fast food outlets. It additionally included height and weight data from state drivers' licenses to compute an average body mass index (BMI) and used epidemiological data of diet related diseases such as diabetes or cardiovascular disease to compute "years of potential life lost" in specific neighborhoods due to poor food access, mapping each statistic at the zip code level (Gallagher 2006, p. 23 ff.).

   While this work shows notably close attention to individual health outcomes, it follows a pattern common in this research of identifying individuals through their places of residence, aggregated to politically defined areal units (census tracts and zip codes in this case). Like the reliance on supermarkets as proxies for healthy food, this decision is understandable. Data on individual movement is difficult to come by.[3] However, this decision fixes urban residents as passive and immobile, ignoring the complex relational and economic factors that influence how individuals gather food for themselves and their loved ones (Valentine 1999; Franzen and C. Smith 2010; Slocum 2007; Zenk, Schulz, Hollis-Neely, et al. 2005; Thornton et al. 2006; Pereira et al. 2010; Wiig and C. Smith 2009). Research on food stamp recipients also supports a more complex perspective: a 1997 study that found "the average distance traveled to redeem SNAP benefits was 2.7 miles, but the average distance to the nearest store was 0.3 miles. These data suggest low-income households typically bypassed nearby supermarkets to use stores farther from home" (USDA Economic Research Service, 2009, p. 63). Residents' food-

---

   3   It is worth noting that a small but growing number of studies also incorporate the use of geographic positioning systems (GPS) and travel diaries to track individual movement to create an analysis of food access based on daily movements (Kestens et al. 2010). As argued below, such studies trace *movement* more than *mobility*, but future studies of food deserts may be improved through the incorporation of such methods.

related mobilities deserve closer attention, including the affective and pragmatic influences that shape food procurement, including the relationships individuals form with particular foods as healthy and/or desirable, coping strategies tied to limited transportation options, individual responsibilities for children and to partners, and the self-regulation of spending habits through the choice of particular retail options (dollar stores over supermarkets, for example).

The word "mobility" deserves some attention here. Cresswell (2006), one of the main scholars identified with current scholarship on the term, contrasts mobility with movement. The latter represents the physical displacement involved in moving individuals from point A to point B. It is "contentless, apparently natural, and devoid of meaning, history, and ideology" (p. 3). Mobility is movement contextualized by meaning and power. It recognizes the social significance of all forms of movement, whether the act of migrating from one country to another, the ability or inability to enter certain spaces, or even the link between identity and styles of movement, such as a way of walking. "Mobile people are never simple people—they are dancers and pedestrians, drivers and athletes, refugees and citizens, tourists or businesspeople, men and women" (p. 4). These social meanings frame movement within a context of power relations, where it is circumscribed by larger political, economic, and social forces. Authors reflecting the so-called "mobility turn" emphasize the affective and power laden dimensions of everyday movement, researching how individual mobilities are impacted by social relations (Cresswell 2010; D. Bell and Hollows 2007; Kabachnik 2009; Urry 2007; Amin and Thrift 2002; Conradson and Latham 2005).

When considering food access, a focus on mobility involves recognition that the forces influencing movement are both deeply personal and social. Understanding neighborhood food access thus means not just tracking *where* people go to get food, but *how* and *why* they go there—how various factors, including vehicle access; gender, race, class, and generational identity; and the characteristics of particular retail sites influence individual behavior. It also highlights the affective nature of spaces of food consumption (Slocum 2007; A. Hayes-Conroy and J. Hayes-Conroy 2008; D. Massey 1994; Probyn 2000; Valentine 1999). Rather than simply analyzing foods' proximity, a focus on mobilities of food consumption highlights the relational and physical networks influencing individuals' food procurement and highlighting how those mobilities are linked to economic and social inequality. This in turn can lead to a more "visceral" food politics which attends to the various forces and appetites which shape individuals' food consumption (A. Hayes-Conroy and J. Hayes-Conroy 2008).

Some examples of this approach already exist. Several studies combine qualitative and quantitative approaches to map urban residents' mobilities, relying on travel diaries, photographs, and mental maps as data sources (Latham 2003; Rogalsky 2010; Matthews et al. 2005; Brennan-Horley 2010). Research done by Hillier et al. (2011) on the shopping habits of WIC recipients is one example of what such an approach might entail when applied to food consumption. Large scale data on consumption behavior, such as records of food stamp purchases or Nielsen's

Homescan data, may be another rich source. The goal of such research would be to highlight both *limits* (self-imposed or environmental) to individual mobility and the ways that individuals travel *beyond* expected geographic boundaries to procure food, as in the case of those who ride along with family or friends to stores outside their neighborhood for cultural or economic reasons (Coveney and O'Dwyer 2009; Franzen and C. Smith 2010). The relational networks enabling this mobility are key considerations in designing better access, particularly as they suggest residents' affective attachments to specific foods and food sites. By loosening the fix on homogenous and immobile urban residents, such research would highlight the variety of food consumption mobilities present even in low-access areas.


**Conclusion**

Lying with maps may be unavoidable, but the purpose of this chapter is to suggest that the three fixes outlined above are more than innocent cartographic choices. These fixes—equating supermarkets and healthy food, homogenizing food territories, and immobilizing urban residents—support an approach that prioritizes environmental influence, downplays the diversity of individual food consumption practices, and reinforces the dominance of large-scale food production and retailing. In pathologizing both these spaces and the individuals that populate them by focusing only on the presence of unhealthy consumption habits and understocked food stores, this research largely fails to consider the political, economic, and social relations which shape both neighborhoods and residents. By applying universal measures of distance, for example, other questions are sidelined: To whom does distance matter? Where, how, and why does it matter? The answers to these questions require a richer and more diverse set of analytical tools than are currently in use.

Certainly, residents of low income neighborhoods experience food insecurity and poor nutrition, in part due to the options available to them. However, by abstracting these problems from their economic and political context and reducing the problem to availability, food desert research sidesteps a myriad of messier, less easily fixed problems: inadequate public transportation networks, processes of urban economic and ethnic segregation, low wages and poor job options, and price supports for commodity crop production. Examination of the food procurement practices of low-income populations would highlight both the effects of these problems and the ways residents have adapted to them, as well as suggesting how struggles over food access may be linked to broader activism around labor issues and local governance.

Should we fix food deserts? Current research has unarguably galvanized political and media attention to disparities in the availability of healthy foods. However, if current efforts to improve the accessibility of healthy food are to succeed, a broader, more adaptable lens on these neighborhoods may prove useful. Michel de Certeau (1984), in a famous analogy, compared modernist urban planners to individuals viewing New York from the quiet roof of a skyscraper. While this vantage point provides a broad view of the city and its structures, it provides little insight on the ways urban spaces are used and traversed by residents. Really addressing the problem of food deserts may well require more research done at street level, amidst the confusion and noise of people moving along multiple pathways to gather and consume their food.

## References

Agrawal, A., 2005. Environmentality: Community, Intimate Government, and the Making of Environmental Subjects in Kumaon, India. *Current Anthropology*, 46(2), pp. 161-190.

Amin, A. and Thrift, N., 2002. *Cities: reimagining the urban*, Cambridge: Polity.

Anon, 2009. Mayor Bloomberg, Governor Paterson and Speaker Quinn announce comprehensive strategies to increase and retain grocery stores in New York City. Available at: http://www.nyc.gov/html/dcp/html/about/pr051609.shtml [accessed December 8, 2009].

Anon, 2010a. Obama Administration Details Healthy Food Financing Initiative. *U.S. Department of Heath and Human Services*. Available at: http://www.hhs.gov/news/press/2010pres/02/20100219a.html [accessed January 25, 2011].

Anon, 2010b. Save-A-Lot Discount Grocery Chain Wants To Expand In Chicago. *CBS2 Chicago*. Available at: http://cbs2chicago.com/local/save.a.lot.2.1462170.html.

Anon, 2011. Pennsylvania Fresh Food Financing Initiative. *The Food Trust*. Available at: http://www.thefoodtrust.org/php/programs/fffi.php [accessed January 31, 2011].

Apparicio, P., Cloutier, M.-S. and Shearmur, R., 2007. The case of Montréal's missing food deserts: Evaluation of accessibility to food supermarkets. *International Journal of Health Geographics*, 6(4), pp.1-13.

Ashbrook, A. and Roberts, K., 2010. When Healthy Food is Out of Reach: An Analysis of the Grocery Gap in the District of Columbia. Available at: www.dchunger.org/pdf/grocerygap.pdf.

Beaulac, J., Kristjansson, E. and Cummins, S., 2009. A systematic review of food deserts, 1966-2007. *Preventing Chronic Disease*, 6(3), p.A105.

Belasco, W., 1993. *Appetite for Change: How the Counterculture Took on the Food Industry*. Updated ed., Ithaca: Cornell University Press.

Bell, D. and Hollows, J., 2007. Mobile Homes. *Space and Culture*, 10(1), pp. 22-39. Available at: http://sac.sagepub.com/cgi/doi/10.1177/1206331206296380 [accessed January 20, 2011].

Black, J.L. et al., 2011. Exploring the distribution of food stores in British Columbia: Associations with neighbourhood socio-demographic factors and urban form. *Health and Place*, 17(4), pp. 961-70.

Boone-Heinonen, J.. et al., 2011. Fast Food Restaurants and Food Stores: Longitudinal Associations With Diet in Young to Middle-aged Adults: The CARDIA Study. *Archives of Internal Medicine*, 171(13), pp. 1162-70.

Bourdieu, P., 1984. *Distinction: a Social Critique of the Judgement of Taste*, Cambridge Mass.: Harvard University Press.

Brennan-Horley, C., 2010. Multiple Work Sites and City-wide Networks: a topological approach to understanding creative work. *Australian Geographer*, 41(1), pp. 39-56.

Brown, M. and Knopp, L., 2006. Places or polygons? Governmentality, scale, and the census in the Gay and Lesbian Atlas. *Population, Space and Place*, 12(4), pp. 223-242.

Byrne, J., 2010. Chicago partners with Walgreens to bring groceries to food deserts. *Chicago Tribune*. Available at: http://articles.chicagotribune.com/2010-08-11/business/ct-biz-0812-food-deserts-20100811_1_food-deserts-wal-mart-grocery-stores [accessed June 10, 2011].

Chomitz, V.R. et al., 2010. Healthy Living Cambridge Kids: a community-based participatory effort to promote healthy weight and fitness. *Obesity (Silver Spring, Md.)*, 18 Suppl 1, pp. S45-53. Available at: http://www.ncbi.nlm.nih.gov/pubmed/20107461 [accessed April 26, 2010].

Clarke, I. et al., 2006. Retail restructuring and consumer choice 1. Long-term local changes in consumer behaviour: Portsmouth, 1980 – 2002. *Environment and Planning A*, 38(1), pp. 25-46.

Clarke, I. et al., 2004. Retail competition and consumer choice: contextualising the "food deserts" debate. *International Journal of Retail & Distribution Management*, 32(0959-0552), pp. 89-99.

Conradson, D. and Latham, A., 2005. Transnational Urbanism: Attending to Everyday Practices and Mobilities. *Journal of Ethnic & Migration Studies*, 31(2), pp. 227-233.

Coveney, J. and O'Dwyer, L. a, 2009. Effects of mobility and location on food access. *Health & Place*, 15(1), pp. 45-55.

Crampton, J.W., 2004. GIS and Geographic Governance: Reconstructing the choropleth map. *Cartographica: The International Journal for Geographic Information and Geovisualization*, 39(1), pp. 41-53.

Crampton, J.W. and Krygier, J., 2006. An Introduction to Critical Cartography. *Cartography*, 4(1), pp. 11-33.

Cresswell, T., 2010. Mobilities I: Catching up. *Progress in Human Geography*, 35(4), pp. 550-558.

Cresswell, T., 2006. *On the Move: Mobility in the Modern Western World*, New York: Routledge.

Cummins, S. and Macintyre, S., 2002. A Systematic Study of an Urban Foodscape: The Price and Availability of Food in Greater Glasgow. *Urban Studies*, 39(11), pp. 2115-2130.

Cummins, S. et al., 2007. Understanding and representing "place" in health research: A relational approach. *Social Science & Medicine*, 65(9), pp. 1825-1838.

De Certeau, M., 1994. *The Practice of Everyday Life*. Berkeley: Univ. of California Press.

Deutsch, T., 2010. *Building a Housewife's Paradise: Gender, Politics, and American Grocery Stores in the Twentieth Century*, Chapel Hill: University of North Carolina Press.

Doe, H. et al., 2011. *10 Ideas for Economic Development*, Washington, D.C.: Roosevelt Institute Campus Network.

Donald, B., 2008. Food Systems Planning and Sustainable Cities and Regions: The Role of the Firm in Sustainable Food Capitalism. *Regional Studies*, 42(9), pp. 1251-1262.

Donofrio, G., 2007. Feeding the City. *Gastronomica*, 7(4), pp. 30-41.

D'Innocenzio, A., 2010. Wal-Mart to aggressively roll out smaller stores. *Salon*. Available at: http://www.salon.com/news/walmart/index.html?story=/news/feature/2010/09/20/wal_mart_urban_expansion.

Egger, G. and Swinburn, B., 1997. An "ecological" approach to the obesity pandemic. *British Medical Journal*, 315, pp. 477-480.

Eicher, C.L. and Brewer, C. a, 2001. Dasymetric Mapping and Areal Interpolation: Implementation and Evaluation. *Cartography and Geographic Information Science*, 28(2), pp. 125-138.

Eisenhauer, E., 2001. In poor health: Supermarket redlining and urban nutrition. *GeoJournal*, 53(2), pp.125-133.

Foucault, M., 1991. Governmentality. In G. Burchell, C. Gordon, and P. Miller, eds. *The Foucault Effect: Studies in Governmentality*. Chicago: University of Chicago Press, pp. 87-104.

Franzen, L. and Smith, C., 2010. Food system access, shopping behavior, and influences on purchasing groceries in adult hmong living in Minnesota. *American Journal of Health Promotion*, 24(6), pp. 396-409.

Gallagher, M., 2006. Examining the Impact of Food Deserts on Public Health in Chicago. , 325(7361). Available at: http://www.ncbi.nlm.nih.gov/pubmed/20462784.

Gallagher, M., 2009. The Chicago Food Desert Update Report. Available at: http://marigallagher.com/site_media/dynamic/project_files/ChicagoFoodDesProg2009a.pdf.

Gibson-Graham, J.K., 1996. *The End Of Capitalism (As We Knew It): A Feminist Critique of Political Economy* 1st ed., Minneapolis: Univ Of Minnesota Press.

Griffioen, J., 2011. Yes, There Are Grocery Stores in Detroit. *The Urbanophile*. Available at: http://www.urbanophile.com/2011/01/25/yes-there-are-grocery-stores-in-detroit-by-james-griffioen/ [accessed January 31, 2011].

Guthman, J., 2011. *Weighing in: Obesity, Food Justice and the Limits of Capitalism*. Berkeley, CA: University of California Press.

Hagerstrand, T., 1982. Diorama, path and project. *Tijdschrift voor economische en sociale*, 73(6), pp. 323-339.

Harvey, D., 2006. *The Limits to Capital*, London: Verso.

Hayes-Conroy, A. and Hayes-Conroy, J., 2008. Taking back taste: feminism, food and visceral politics. *Gender, Place & Culture: A Journal of Feminist Geography*, 15(5), pp. 461-473.

Hemphill, E. et al., 2008. Exploring obesogenic food environments in Edmonton, Canada: the association between socioeconomic factors and fast-food outlet access. *American Journal of Health Promotion: AJHP*, 22(6), pp. 426-432.

Hillier, A. et al., 2011. How Far Do Low-Income Parents Travel to Shop for Food? Empirical Evidence from Two Urban Neighborhoods. *Urban Geography*, 32(5), pp. 712-729.

Jones, R., Pykett, J. and Whitehead, M., 2010. Governing temptation: Changing behaviour in an age of libertarian paternalism. *Progress in Human Geography*, 35(4), pp. 483-501.

Jones, S.M., 2011. Bull's-eye for Target, city. *Chicago Tribune*. Available at: http://articles.chicagotribune.com/2011-02-16/business/ct-biz-0216-target-loop-20110216_1_target-plans-carsons-retailer [accessed March 7, 2011].

Kabachnik, P., 2009. To choose, fix, or ignore culture? The cultural politics of Gypsy and Traveler mobility in England. *Social & Cultural Geography*, 10(4), pp. 461-479.

Kestens, Y. et al., 2010. Using experienced activity spaces to measure foodscape exposure. *Health & Place*, 16(6), pp. 1094-103.

Larsen, K. and Gilliland, J., 2008. Mapping the evolution of "food deserts" in a Canadian city: Supermarket accessibility in London, Ontario, 1961–2005. *International Journal of Health Geographics*, 7, p.16.

Latham, A., 2003. Urbanity, Lifestyle and Making Sense of the New Urban Cultural Economy: Notes from Auckland, New Zealand. *Urban Studies*, 40(9), p.1699.

Lawson, L.J., 2005. *City Bountiful: A Century of Community Gardening in America*, Berkeley: University of California Press.

Lee, G. and Lim, H., 2009. A Spatial Statistical Approach to Identifying Areas with Poor Access to Grocery Foods in the City of Buffalo, New York. *Urban Studies*, 46(7), pp. 1299-1315.

Levenstein, H., 1988. *Revolution at the Table: The Transformation of the American Diet*, Oxford: Oxford UP.

Luke, T.W., 1999. Envronmentality and Green Governmentality. In É. Darier, ed. Wiley-Blackwell.

Luke, T.W., 1995. Sustainable development as a power/knowledge system: the problem of "governmentality." In F. Fischer and M. Black, eds. *Greening Environmental Policy: Politics of a Sustainable Future*. London: Palgrave Macmillan.

Massey, D., 1994. *Space, Place and Gender*, Minneapolis: University of Minnesota Press.

Massey, D.B., 2005. *For Space*, London: Sage Publications Ltd.

Matthews, S.A., Detwiler, J.E. and Burton, L.M., 2005. Geo-ethnography: Coupling Geographic Information Analysis Techniques with Ethnographic Methods in Urban Research. *Cartographica*, 40(4), pp. 75-90.

McClintock, N., 2011. From Industrial Garden to Food Desert: Demarcated Devaluation in the Flatlands of Oakland, California. In A. Alkon and J. Agyeman, eds. *Cultivating Food Justice: Race, Class, and Sustainability*. Cambridge MA: MIT Press, pp. 135-179.

McPhail, D., Chapman, G.E. and Beagan, B.L., 2011. "Too much of that stuff can't be good": Canadian teens, morality, and fast food consumption. *Social science & medicine*, 73(2), pp. 301-7.

Mennis, J., 2003. Generating surface models of population using dasymetric mapping. *Professional Geographer*, 55(1), pp. 31-42.

Mertens, A., 2011. The West Loop is not a food desert. *Chicago Journal*. Available at: http://www.chicagojournal.com/News/08-10-2011/The_West_Loop_is_not_a_food_desert [accessed August 11, 2011].

Mitchell, G., 2004. Libertarian paternalism is an oxymoron. *Nw. UL Rev.*, 99(3), p.1245.

Monmonier, M., 1991. *How to Lie with Maps*, Chicago: University of Chicago Press.

Odoms-Young, A.M., Zenk, S.N. and Mason, M., 2009. Measuring food availability and access in African-American communities: implications for intervention and policy. *American Journal of Preventive Medicine*, 36(4 Suppl), pp. S145-50.

Ogburn, S., 2010. Would a Walmart solve West Oakland's and Nashville's food problems? *Grist*. Available at: http://www.grist.org/article/food-2010-10-05-would-a-walmart-solve-oaklands-and-nashvilles-food-problems/ [accessed October 8, 2010].

Pavlovskaya, M., 2004. Other Transitions: Multiple Economies of Moscow Households in the 1990s. *Annals of the Association of American Geographers*, 94(2), pp. 329-351.

Pereira, C.A.N., Larder, N. and Somerset, S., 2010. Food acquisition habits in a group of African refugees recently settled in Australia. *Health & Place*, 16(5), pp. 934-41.

Poulsen, E. and Kennedy, L.W., 2004. Using dasymetric mapping for spatially aggregated crime data. *Journal of Quantitative Criminology*, 20(3), pp. 243-262.

Pred, A., 1984. Place as Historically Contingent Process: Structuration and the Time-Geography of Becoming Places. *Annals of the Association of American Geographers*, 74(2), pp. 279-297.

Probyn, E., 2000. *Carnal Appetites: FoodSexIdentities* 1st ed., Routledge.

Raco, M. and Imrie, R., 2000. Governmentality and rights and responsibilities in urban policy. *Environment and Planning A*, 32(12), pp. 2187-2204.

Raja, S. and Yadav, P., 2008. Beyond Food Deserts: Measuring and Mapping Racial Disparities in Neighborhood Food Environments. *Journal of Planning Education and Research*, 27(4), pp. 469-482.

Raja, Samina, Yadav, Pavan and Ma, C., 2008. Beyond Food Deserts: Measuring and Mapping Racial Disparities in Neighborhood Food Environments. *Journal of Planning Education and Research*, 27(4), pp. 469-482.

Richards, E., 1893. Scientific Cooking Studies in the New England Kitchen. *The Forum*, pp. 356-7.

Rogalsky, J., 2010. The working poor and what GIS reveals about the possibilities of public transit. *Journal of Transport Geography*, 18(2), pp. 226-237.

Rutherford, S., 2007. Green governmentality: insights and opportunities in the study of nature's rule. *Progress in Human Geography*, 31(3), pp. 291-307.

Schuurman, N. et al., 2006. Building an integrated cadastral fabric for higher resolution socioeconomic spatial data analysis. In *Progress in Spatial Data Handling*. Berlin: Springer Berlin Heidelberg, pp. 897-920.

Scott, J.C., 1998. *Seeing Like a State: How Certain Schemes to Improve the Human Condition have Failed*, New Haven: Yale University Press.

Shapiro, L., 1986. *Perfection Salad: Women and Cooking at the Turn of the Century*, New York: Farrar Straus and Giroux.

Shaw, H.J., 2006. Food Deserts: Towards the Development of a Classification. *Geografiska Annaler, Series B: Human Geography*, 88(2), pp. 231-247.

Short, A., Guthman, J. and Raskin, S., 2007. Food Deserts, Oases, or Mirages?: Small Markets and Community Food Security in the San Francisco Bay Area. *Journal of Planning Education and Research*, 26(3), pp. 352-364.

Short, Anne, Guthman, J. and Raskin, Samuel, 2007. Food Deserts, Oases, or Mirages?: Small Markets and Community Food Security in the San Francisco Bay Area. *Journal of Planning Education and Research*, 26(3), pp. 352-364.

Simon, H.A., 1955. A behavioral model of rational choice. *The quarterly journal of economics*, 69(1), pp. 99-118.

Slocum, R., 2007. Whiteness, space and alternative food practice. *Geoforum*, 38(3), pp. 520-533.

Smith, J. and Jehlička, P., 2007. Stories around food, politics and change in Poland and the Czech Republic. *Transactions of the Institute of British Geographers*, 32(3), pp. 395-410.

Sparks, A., Bania, N. and Leete, L., 2009. Finding Food Deserts: Methodology and Measurement of Food Access in Portland, Oregon. In *Institute of Medicine, Workshop on the Public Health Effects of Food Deserts*. January.

Stephens, J., 2007. Los Angeles Tries to Zone Out Fast Food. *Planning*, 73(11), p. 48.

Stokols, D., 1995. Translating Social Ecological Theory into Guidelines for Community Health Promotion. *American Journal of Health Promotion*, 10(4), pp. 282-298.

Strauss, K., 2007. Re-engaging with rationality in economic geography: behavioural approaches and the importance of context in decision-making. *Journal of Economic Geography*, 8(2), pp. 137-156.

Swinburn, B., Egger, G. and Raza, F., 1999. Dissecting Obesogenic Environments: The Development and Application of a Framework for Identifying and Prioritizing Environmental Interventions for Obesity. *Preventive Medicine*, 29(6), pp. 563-570.

Tangires, H., 2002. *Public markets and civic culture in nineteenth-century America*, Baltimore: Johns Hopkins University Press.

Thaler, R.H. and Sunstein, C.R., 2003. Libertarian Paternalism. *American Economic Review*, 93(2), pp. 175-179.

Thaler, R.H. and Sunstein, C.R., 2008. *Nudge: Improving Decisions about Health, Wealth and Happiness*, Cornwall: Yale University Press.

Thornton, P.L. et al., 2006. Weight, diet, and physical activity-related beliefs and practices among pregnant and postpartum Latino women: the role of social support. *Maternal and Child Health Journal*, 10(1), pp. 95-104.

Tversky, A. and Kahneman, D., 1974. Judgment under Uncertainty: Heuristics and Biases. *Science*, 185(4157), pp. 1124-31.

USDA Economic Research Service, 2009. Access to Affordable and Nutritious Food—Measuring and Understanding Food Deserts and Their Consequences: Report to Congress.

Urry, J., 2007. *Mobilities*, Cambridge: Polity.

Valentine, G., 1999. A Corporeal Geography of Consumption. *Environment and Planning D: Society and Space*, 17(3), pp. 329-351.

Walker, K.E., 2010. Negotiating GIS and Social Theory in Population Geography. *Geography Compass*, 4(6), pp. 616-629.

WalmartCommunity, 2010. Chicago residents talk about Walmart. Available at: http://www.youtube.com/watch?v=uFaGNhkDrXQ&list=FLukDukJRgfJejQe e6phwWvQ&index=15 [accessed August 19, 2011].

Warren, J., 2010. How Wal-Mart Won Chicago. *BusinessWeek*. Available at: http://www.businessweek.com/magazine/content/10_30/b4188037407551.htm. [accessed August 10, 2011].

Watts, M., 2002. Preface: Green Capitalism, Green Governmentality. *American Behavioral Scientist*, 45(9), pp. 1313-1317.

Whelan, A. et al., 2002. Life in a "Food Desert." *Urban Studies*, 39(11), pp. 2083-2100.

Wiig, K. and Smith, C., 2009. Factors Affecting Low-income Women's Food Choices and the Perceived Impact of Dietary Intake and Socioeconomic Status on Their Health and Weight. *Journal of Nutrition Education & Behavior*, 41(4), pp. 242-253.

Wood, D. and Fels, J., 1992. *The Power of Maps*. New York: Guilford.

Wood, D., Fels, J. and Krygier, J., 2010. *Rethinking the Power of Maps*, New York: Guilford Press.

Wrigley, N., 2002. "Food Deserts" in British Cities: Policy Context and Research Priorities. *Urban Studies*, 39(11), pp. 2029-2040.

Zenk, S.N., Schulz, A.J. and Israel, B., 2005. Neighborhood racial composition, neighborhood poverty, and the spatial accessibility of supermarkets in metropolitan Detroit. *Journal of Public*, 95(4), pp. 660-667.

Zenk, S.N., Schulz, A.J., Hollis-Neely, T., et al., 2005. Fruit and vegetable intake in African Americans income and store characteristics. *American Journal of Preventive Medicine*, 29(1), pp. 1-9.

# Thematic Tabs for Chapter 15

Body

Discourse

Structure

Women

Take-Home Labs for Chapter 13

Chapter 15

# Mobilizing Caring Citizenship and Jamie Oliver's Food Revolution

Heidi Zimmerman

**Editors' Note**: This chapter relates to four themes: *body, discourse, structure* and *women*. In the vein of scholars like Julie Guthman, Patricia Allen and Melanie Dupuis (e.g. Guthman and Dupuis 2006, Allen and Guthman 2006) this chapter draws upon critiques of neoliberalism to analyze the popular television series, Jamie Oliver's Food Revolution. In doing so, Zimmerman works to counter the belief that obesity is largely a matter of personal responsibility, and contributes to ongoing scholarly and activist debates that seek to change the ways in which body size, overweight and obesity are approached and understood. One of the central contributions of this chapter to doing nutrition differently is to insist that the seemingly individual 'problem' of nutrition must be understood from within the wider socio-political and cultural contexts in which each consumer is embedded.

## Introduction

Between March and April of 2010, British celebrity chef Jamie Oliver invited television viewers to join him as he staged a "food revolution" in the public schools of Huntington, West Virginia. The "revolution" is chronicled in a six-episode ABC reality series entitled *Jamie Oliver's Food Revolution*. Huntington, the show argues, has special need for revolutionary intervention because it was recently declared "the unhealthiest city in America" based on a Center of Disease Control report which cited the rates of obesity, among other factors. The series follows Oliver as he attempts to shift lunchtime fare in the schools' cafeterias from processed foods to meals made from scratch, out of fresh ingredients, and crucially, "with love." In other words, the nutrition intervention that the show brings to Huntington is not only about ideas of "good food," but also about particular ways of caring.

The show engages viewers in caring about Huntington by representing it as at once normatively ordinary and gravely endangered. The city functions as a kind of test site to demonstrate the possibilities of this particular "food revolution" to solve what is perceived to be a national crisis of diet and nutrition. For example, in the opening of the premiere episode, a male voiceover explains we are entering "Beautiful Huntington, West Virginia. Population fifty thousand. Home of Marshall University." During this introduction, we are treated to views of university buildings and tree-lined streets in apparently thriving commercial districts, all peopled by students and happy-looking shoppers, not particularly marked by fatness. This montage constructs Huntington as normatively "typical"—and implies that "typicality" is marked by whiteness, thinness, and access to higher education and consumer pleasures. As the visuals shift to low-angle medium shots of headless fat bodies, ostensibly walking around Huntington, the voiceover continues: "And recently named ... *The unhealthiest city in America.*" Viewers receive spectacular and shocking evidence of this ill-health from multiple sources: Pastor Steve Willis pages through a church directory, noting each person who died prematurely from "complications from obesity;" a local funeral home displays an extra-wide casket. The narrator continues: "In a place where nearly half of the adults are considered obese and incidence of heart disease and diabetes lead the nation, one man will try to save 50 thousand lives."

*Jamie Oliver's Food Revolution* (*JOFR*) thus introduces a broad problem of "obesity" and represents it as a particularly urgent one. Once this problem was established, the show could have gone in any of several directions. It might, for example, have explored the ways in which historical processes that have produced high rates of poverty overlap with processes that have contributed to high rates of obesity—indeed, in 2008, when the CDC report was published, the poverty rate in Huntington was above the national average (Stobbe 2008). It could have explored the relationship of state and federal budget decisions to the nutritional value of school lunches. Or it could have investigated the political economy of school lunches in the context of the industrial food system and present-day capitalism, that is, the circuit of production, distribution, and consumption of school lunches, who profits from this system, and who bears the costs. However, *JOFR* does none of these things. Instead, as I will demonstrate, for *JOFR*, the problem of obesity is perpetuated not by poverty, inadequate State investment, or private-sector interests, but rather by individual failure to care.

Thus, to understand *JOFR* as merely a nutrition intervention, and to ignore the role it assigns to what I call caring citizens, risks failing to see its broader political implications. The narrative coherence of *JOFR*, like many nutrition interventions that appear in popular media, relies on the assumption that the solution to what had been called the "obesity epidemic" is a matter of personal responsibility and individual free choice—that is, if individuals responsibly choose nutritious foods for themselves and for those whom they feed, Huntington (and, implicitly, America) would not have such a problem. Yet, ironically, the show cannot help but demonstrate the impossibility of such a solution—especially given the fact that its sites of intervention are the public schools, which are hopelessly entangled with larger structural questions of the State and food industry contracts. The problem of obesity is far beyond the scope of individual free choice. In other words, individuals' lives are structured by geographic, time, material, financial, familial, and institutional constraints that put limits on the "freedom" of their choices. Additionally, the marketplace within which individuals operate is far from free.

It is highly structured by powerful industries motivated by profit, not health or justice, which provide only limited choices for consumers.

It is also worth pointing out that the show takes for granted that fatness itself is a problem. Julie Guthman has persuasively argued that the link between obesity and mortality is far too complex and uncertain to read the risk of early death off of fat bodies themselves. Particularly, Guthman surveys relevant epidemiological literature to point to the limitations of using the measure of fatness determined by BMI (body mass index, which calculates one's weight to height ratio) to determine the risk of developing non-communicable diseases like heart disease and diabetes (Guthman 2011). Images of fat bodies, as used in *JOFR,* are even more imprecise and arbitrary than the BMI measure. More distressing still, the impact of such images relies upon normative notions of body aesthetics and the fact that publically ridiculing, objectifying, and scolding people for the fatness of their bodies remains culturally acceptable.

To place personal responsibility at the center of efforts to fight obesity not only functions to blame individuals for massive and complex structural conditions, it also elides the role of the State and of private industry in the problem itself. This individualized problematization of obesity shuts off solutions that might address, instead: the structures that produce poverty, low wages, or unemployment; those that make health care inaccessible to many Americans; or the reasons why unhealthful foods are produced and distributed in the ways that they are.

*JOFR* attempts to argue that individual choice-making is an adequate solution to Huntington's purported ill-health. The narrative drama is structured around Oliver's efforts to cultivate caring investment among cafeteria workers, parents, children, within private industry, and even among individual State actors. Yet at each site of intervention, contradictions threaten to disrupt this narrative of individual responsibility. By contradictions, I mean moments within the show that gesture to the reality that individual responsibility is an insufficient response to ill-health and bad cafeteria food, moments that suggest that structural changes are necessary. I argue that the show mobilizes the notion of care precisely to manage the contradictions that emerge when a project ostensibly aimed at food justice fails to challenge the broader cultural fiction that health and success are secured through personal responsibility and choice exercised within an ostensibly "free" market, a fiction central to what many scholars have called neoliberalism.

## Neoliberalism and Problems of Citizenship

I understand neoliberalism as the turn toward what David Harvey describes as "Deregulation, privatization, and withdrawal of the state from many areas of social provision" that began in the 1970s and was aggressively advanced by the policies of Reagan administration in the US and those of Margaret Thatcher in the UK (Harvey 2005, 3). Neoliberalism entails a state apparatus whose primary role is to ensure the "proper functioning of markets" (Harvey 2005, 2). This kind of state is often opposed to a welfare state apparatus, which provides more extensive public safety net for citizens. As the state has rolled back the social safety net, television, government programs, health care plans and lifestyle literature have increasingly called upon individuals to take responsibility for their own wellbeing, so the state does not have to (Ouellette and Hay 2008). Although this shift might *appear* to be value-neutral or amoral, Wendy Brown has pointed out that the manner in which neoliberalism figures the individual has a very specific moral bent:

It figures individuals as rational, calculating creatures whose moral autonomy is measured in their capacity for 'self-care'—the ability to provide for their own needs and service their own ambitions. In making the individual fully responsible for him- or herself, neoliberalism equates moral responsibility with rational action; it erases the discrepancy between economic and moral behavior by configuring morality as a matter of rational deliberation about costs, benefits, and consequences. But in doing so, it carries responsibility for the self to new heights: the rationally calculating individual bears full responsibility for the consequences of his or her action no matter how severe the constraints on this action—for example, lack of skills, education, and child care in a period of high unemployment and limited welfare benefits (Brown 2005, 42-43).

In other words, the shift toward financialization, monetization, and privatization that characterizes neoliberalism also entails a shift in values. For Brown, if neoliberalism could talk it would say this: irrespective of any material, economic, cultural, social, or geographic constraints you might face, if you fail to thrive (that is, if you get sick, get fat, can't work, can't afford childcare, lose your job or your home) within present-day economic conditions, *it's your fault because you have failed to make the right choices.*

This central tenet of neoliberalism is highly contradictory, of course. The "self-care" to which Brown refers is imagined to be a question of "free choice." Yet caring for the self is rarely a matter of "free choice" and, further, no matter how well executed, such care does not guarantee that an individual will be able to succeed. Present-day social, political and economic structures make "free choice" and success a near impossibility for a great many people, in ways that are racialized, gendered and classed. Those who seem to succeed by dint of their own stick-to-itiveness, in fact benefit from a great many structural advantages that undermine the notion that pulling one's self up by one's bootstraps is possible at all. If we imagine that individual self-care is an adequate solution to structural problems (like unemployment, poverty, or the inability of the healthcare system to accommodate the healthcare needs of the people), there will remain a gap between what individuals can realistically do for themselves, and what they require to survive, much less thrive. When this gap produces social crises, like the "obesity epidemic," the neoliberal state does not sponsor social services or invest in school lunch programs within the public school system. Rather, the state makes way for private sector initiatives and public-private partnerships, which increasingly offer stopgap measures at the least possible cost to the state. Although proponents of neoliberal thinking maintain that intensifying privatization can alleviate such crises, I argue, on the contrary, that private solutions function to manage the contradictions within neoliberalism so as to justify the maintenance of the status quo and make structural change less thinkable.

## Problematization of Food/Feeders in *JOFR*

In the following section of this chapter, I will discuss the ways in which *JOFR* advances the argument that real power to effect change lies in the caring investment of individuals actors. That is, the central problem with which the show is concerned is individual failures to care. The narrative arc of *JOFR* depends upon showcasing such failures to care, then enrolling individuals in a training process geared toward empowered, caring citizenship, and, finally, celebrating individual performances of care as ends in themselves. Although I will focus mainly on the ways in which this set-up individualizes the problem of bad food and disarticulates it from larger structures, in tracing this argument I will also point out the moments at which the limitations of *JOFR*'s solution break through the narrative and, perhaps, offer possibilities for imagining other forms of intervention.

*JOFR* emphasizes the "don'ts" of caring citizenship most explicitly among cafeteria workers. Cafeteria workers are a convenient site at which to demonstrate these don'ts because the show addresses viewers as parents, rather than workers. This strategy allows the show to problematize bad attitudes while distancing this problem from viewers. Despite the fact that the cooks have little institutional power, the show points to their hostility as the central obstacle to effecting change. For example, cafeteria cook, Alice Gue registers her irritation: "I don't understand why he's here to change our system, which is working good." The narrative positions Gue as an obstacle to the realization of Oliver's vision, a "force to be reckoned with." Thus, Gue's cooperation becomes necessary for the solution—the show suggests that the success or failure of experiments like Oliver's depends on the Alice Gues of the world—and their willingness to put in "extra effort" at work. The show heightens the stakes of "positive attitude" at the expense of investigating the broader system in which Gue operates. Oliver steadfastly maintains the central importance of good attitudes on the part of cooks, even when this solution is explicitly contradicted. For example when Gue tells Oliver about the challenges the workers have faced in their attempts to implement his recommendations—they have bumped up against the constraints of time and labor. Here, Gue's difficulties might suggest that appropriate action might involve collective demands aimed at the district to hire more cooks or provide additional resources for the existing cooks. Yet, for Oliver, the difficulties, rather, constitute a failure to care. For him, Gue is "blocking" the whole revolution. "I need to get her to care," he resolves. Thus, for the show, the recalcitrance of individual citizens is the central obstacle to positive social change and, as such, overcoming this obstacle by getting individuals to care is the obvious solution. Oliver's invocation of care simplifies and deflects Gue's objections by moralizing her refusal as bad and recentralizing the importance of individual caring (*JOFR* 1.1 2010).

The show thus equates caring with empowerment and obligates this caring as a requirement of good citizenship. It asserts this equation regardless of whether any institutional, financial, or structural changes have taken place. This equation erases both the structural obstacles to making healthful foods in the cafeteria and, as far as a good worker must be a caring worker, obligates additional labor on the part of the cafeteria cooks, apparently with no additional pay. Oliver tells viewers that, in absence of other supports, it is the cultivation of care in the cafeteria cooks that will allow them to overcome the system: "These lunch ladies, they *can* do anything, but they need the permission and the power from above to tell them to do it. But at the same time, I want these lunch ladies to *care* about what's in the food that they're cooking for the kids." This implausible solution to a massive structural problem is pointed out by another of the cafeteria workers: "you're taking this up with us when you should be taking this up with the one who is over us...We can't go in ourselves and fix it. It has to come from the top—" Rather than take her concerns seriously, Oliver interrupts, "and I know that feeling." Oliver implies that this is merely a defeatist attitude, not an informed analysis. He dismisses the relevance of systemic realities as "feeling" and suggests that it is not larger structures that ought to be changed (for example, the number of cooks employed and the autonomy and resources available to them), but rather the *attitudes* of the cooks themselves. Showcasing the negative attitudes in this way serves a dual purpose. On the one hand, it demonstrates the "don'ts" of caring citizenship for viewers to guard against and, on the other hand, it devolves responsibility—not only for remedying the poor nutrition of school lunches, but for solving childhood obesity more broadly—to individual workers in the cafeteria. Although cooks register objections, the moral weight of the obligation to care disallows sustained attention to the systemic issues not remedied by additional effort on the part of individual cooks (*JOFR* 1.2 2010).

**Disgusting Food and Bad Feeding**

*JOFR* chiefly addresses viewers as parents, whose moral obligation to care has broad cultural resonance. While the representation of cafeteria workers helps to establish the threat of negativity (which, perhaps, could also be described as ambivalence about taking on extra labor without institutional support or compensation), the representation of parents functions to shore up the notion that the problem of bad food/feeding is located not in large structures, but in individuals, and in turn to emphasize the possibilities of individual caring for solving the this problem. In so doing, the show uses spectacle to construct certain kinds of feeding and certain kinds of food as self-evidently disgusting. I use the term spectacle to refer to images designed for high visual and emotional impact, usually depicting something the show deems "excessive." The images depict objects removed from their everyday context and are given new meaning through their use within the narrative of the show. Coming to feel shocked and disgusted by "bad" feeding and "bad" food is the process of becoming a good parent and caring citizen—both for viewers at home and for the parents on the show. This shock and disgust tend to locate the *source* of the problem (of obesity and/or poor nutrition) in either the parent her- or him- (but usually her) self or as inherent in the "bad" foods, rather than in the structures in which these parents are operating (like poverty or single motherhood, for example) or in the systems that bring food to the schools.

*JOFR* problematizes bad feeding by entering the home of the Edwards family and exposing its failures on national television. The Edwards are all overweight; the youngest one is pre-diabetic. Mom, Stacey Edwards, walks Oliver through her daily food routine—making chocolate-glazed donuts, frying her own tortilla chips, etc. Pointing to the deep-fryer, Oliver asks, "This is the most highly-used bit of kit in the kitchen, isn't it really?" Stacey responds, "Yeah… I know it's not good for me." Oliver piles a week's-worth of the family's food on the kitchen table: bowls of fries and nuggets, piles of frozen and delivered pizzas, a mountain of corndogs, a plate of burgers stacked high, next to an identical one of hotdogs. Pointing to this excess, Oliver implores Edwards to care, "I need you to know that this is going to kill your children early." For the viewers and for Edwards, the spectacle of the food constructs both the badness of the food itself and the badness of the ways in which Edwards feeds her family. The show argues that the source of childhood obesity lies within Edwards (and others like her) and within the foods she provisions. The conditions that she is up against—the expense of healthcare, the added time, labor and expense of shopping for, cooking and serving healthful foods, the taste proclivities of husband and sons, the fact that her husband, a truck driver, is often away from home, her socio-economic realities—are not figured as obstacles to Edwards's ability to *freely choose* the foods and practices that Oliver recommends. Rather, if she does not choose these practices, it will be *she* who is responsible for early deaths of her family members.

Continuing in this line of reasoning, *JOFR* calls upon other parents in the show, as well as viewers at home, to internalize responsibility for bad food and feeding by shocking them with the spectacle of bad food. He conducts an "experiment" in which he pours out gallons of chocolate milk, buckets of sloppy joe filling, french fries, nachos—"just a months-worth of meals… what's killing our kids and yours." A dump truck tips out great hunks of fat into a dumpster, an amount Oliver tells us is equivalent to that in a year's-worth of cafeteria food and "one of the biggest problems that's immediately affecting our kids."

"Are you parents fine with this?" Oliver demands, standing atop the fat.

"No!"

Yet there is little that is shocking about the food *per se*. It is made shocking *through* the spectacle. For Oliver, seeing all the "crap ingredients" and processed food "pisses me off. And if you're a parent, it should piss you off." The mode of address thus calls upon viewers to identify with the parents, and the "good" or correct moral position is made abundantly clear: a parent who cares ought to be outraged by the fat, sugar and processed ingredients in cafeteria food. A failure to be shocked, a failure to find the food disgusting, constitutes a failure to adequately care. Through its representation of the parents *JOFR* again shores up the equation of caring with empowerment. Oliver emphasizes the exciting possibilities represented by pissed off parents, who "are like a nuclear weapon... if you really upset the parents, you can have everything you want." The show, however, does not articulate what form of change this empowerment ought to be directed toward. Rather, the outrage of the parents is represented as an end in itself (*JOFR* 1.2 2010).

Yet to a limited extent, the show cannot sustain the myth that "empowering" parents like Edwards and viewers at home with knowledge and care is a sufficient solution. At intervals, more coherent critiques of broader political and economic structures are allowed to slip in to the narrative. For example, when Oliver checks on the Edwards family and discovers non-compliance with the menu he assigned to them, he takes the whole family to the doctor. The doctor delivers frightening diagnoses. Although at first, this segment appears to reinforce the notion that one's health is solely a matter of personal responsibility, in *JOFR*, a broader critique creeps in: "Look, I don't know what's happened in this country with health care, I don't understand it, but all I know is I think it's shocking, scary and strange that this family requires me to take them to the hospital to find out what's going on." Here, Oliver suggests this problem might not in fact be due to the failure of individuals to exercise choice responsibly, that it might indicate a broader failure of the US healthcare system to provide an adequate safety net for citizens. Nevertheless, after the briefest of moments, this suggestion is abandoned. It rests uncomfortably in a narrative that favors the possibilities of individually caring citizens (*JOFR* 1.2 2010).

The narrative arc of the show thus sets up the "problems" of hostility, bad parenting, and bad foods and celebrates the way in which the cafeteria workers and parents come to care (in the manner favored by the show) over the course of the series. The pleasures in following the show are thus bound to viewer investment in the transformative possibilities of caring citizenship.[1] The show calls upon viewers to cultivate caring investment within themselves while demonstrating a process of becoming-caring through those on the show. *JOFR* suggests that caring citizenship is an end in itself. Through coming to care, through choosing appropriate caring practices (that is, through creating more nutritious foods at home and in the schools, which involves increased labor without additional resources) it is the responsibility of parents and cafeteria workers to solve the problem of obesity in Huntington.

---

1 Of course, one cannot assume viewers will watch and take pleasure in a show in the manner favored by the narrative. Alternative meanings and pleasures are always at play.

Yet, of course, the increased level of individual caring—which is by definition an increased willingness to invest one's own labor and time to make up for a larger failure to adequately fund school lunch programs (as well as to make up for gross inequalities in healthcare access)—is not sufficient to effect the changes that Jamie Oliver envisions: a total revamping of the food offerings in the cafeteria. The "revolution" needs money, State support, and it needs the kids to eat the food; none of these needs are met through caring efforts. Caring citizenship on the part of parents and cafeteria workers fits neatly within broader notions of neoliberal citizenship, thereby perpetuating the myth that the responsible exercise of free choice guarantees health and success. However, because the show operates within the *public* school and focuses on those with little access to power (particularly the school children), the show is also forced to confront the realities that "free choice" and "personal responsibility" are not, in fact, sufficient when one is up against conditions of inequality and resource scarcity, in which choices are not made in the context of freedom. However, unsurprisingly, the show does not take up this contradiction to launch a critique of economic and political systems that perpetuate poverty and fail to fund school lunch programs. Rather, as I will discuss, the show attempts to contain these contradictions by attempting to pull children, private industry, and State into the logic of caring citizenship.

## School Lunch as Citizen Training: Problems of Choice and Freedom

When Oliver enters the school cafeterias, his central challenge is that he must get the kids to "accept the food." If the kids don't like the food, Oliver will not be given license to expand the program. The problem, here, is whether kids are willing to eat the more nutritious foods that Oliver prepares. *JOFR* thus sets up narrative drama that depends upon conceiving of children as choice-making consumers in a marketplace of cafeteria food options. On the one hand, this is problematic for the obvious reason that children cannot be counted upon to exercise choice responsibly. On the other hand, however, this is problematic because it suggests that free choice on the part of children (rather than increased State funding of school lunches, for example) ought to be a solution to childhood obesity. In order to make this dubious argument plausible, the show launches a citizen-training program for the children in order to get them to *care* about food and nutrition.

*JOFR* assumes that the central obstacle to children's willingness to exercise food-choice responsibly is their lack of knowledge about food. Knowledge, the show suggests, will get them to care, and in turn, make good choices. However, children's behavior, of course, is often not motivated by rational choice-making. Thus, investing in the possibilities of children's informed free choice is risky business. This illogic comes to the fore in Oliver's never-fail "chicken nugget experiment." Oliver tells us, "I'm going to do something really extreme with these kids and get them to *care* what goes into their bodies." He makes chicken nuggets "from scratch" before a group of elementary school students. While Oliver pulverizes chicken carcasses with salt, flavorings and stabilizers, the kids emit shrieks of grossed-out delight. The experiment falls flat when Oliver, holding up a completed nugget asks, "now who would still eat this?" Nearly every hand in the room shoots into the air. The students' responses to Oliver's subsequent questions indicate that they *know* that the food is "bad ... disgusting and gross," yet they delight in eating it anyway (*JOFR* 1.2 2010). In a second attempt to cultivate care in children, Oliver brings a basket of vegetables into a 1ˢᵗ grade classroom. The students cannot identify a single one of the vegetables. Nuggets, fries, and burgers, on the other hand, are correctly named immediately. Oliver bemoans this lack of knowledge as a major obstacle to whether he can get them to accept the food he makes, and thus the success of the program in general. Like the spectacles of "bad foods" and "bad feeding" discussed earlier, the children's failed performances are showcased to elicit the viewers' shock. The segments gesture toward parents and teachers off-screen who have failed to inculcate forms of knowledge and care in the kids, and are thus the implicit targets of the shock, and also the implicit obstacles to a healthier Huntington (*JOFR* 1.2 2010).

Kids themselves (and, by implication, their care-givers/teachers), are thus represented as a *source* of the problem of bad nutrition, thanks to their failure to know and care about food. Yet like the parents, kids are represented as a powerful voting bloc and thus responsible for solving this problem, for, as Oliver complains, "if the kids don't eat [my meals], there's no way [the county food service director is] going to let me back into the school to cook again." In this way, the kids, in their constrained freedom, have the ultimate power. Their vote, the show implies, is the deciding one. If the kids are so potentially empowered, the stakes of training them in appropriate food citizenship—defined as knowing about food, caring (in a particular way) about what they put in their bodies—become very high indeed. They must be taught to exercise this power responsibly for the health of Huntington rests on their choices. The show thus comes dangerously close to blaming the taste proclivities of children for the bad food in cafeterias while simultaneously shoring up the argument that it is in the exercise of individual choice, rather than in collective demands, that real power lies.

Yet, the show cannot sustain the myth that *children* can be counted on to make choices responsibly. These citizens-in-training lack the resources to conduct themselves freely and responsibly in the cafeteria. Trays upon trays of Oliver's "from scratch" fare end up in the trash because the kids have opted not to eat it. Kids continue to choose the sugary, flavored milk over the plain. The show is forced to abandon its investment in the possibilities of kids' individual choices: "For these kids at this age," Oliver concedes, "having choice is probably not a good thing." The next day there was "one choice" and it was Oliver's choice. Although the removal of choice might seem to undermine the overarching argument of the show, the logic of individual caring is immediately reasserted: the show highlights the fact that teachers and the principal have taken it upon themselves to circulate in the cafeteria, prodding students to eat Oliver's food. An especially exemplary teacher designs a lesson plan around vegetable identification. These public school employees take on work beyond their job descriptions to demonstrate that they care. This is not to say that nutrition education for children is undesirable or that teachers and principals should not care about their students. However, given that many teachers—especially in underfunded school districts—already find themselves stretched thin, supplementing scarce resources from their own pocketbooks and finding that they have little freedom to design creative curriculum in their efforts to meet rigid state standards, it seems likely that this teacher created and taught the nutrition lesson as an act of care above and beyond that for which the district compensates her. The show does not suggest that teachers ought to demand compensation or resources for such extra work. Quite the contrary, it naturalizes the notion that, in order to protect children from bad nutrition, public school staff is morally obligated to carry out uncompensated caring work to help the children become responsible choice-makers in the cafeteria.

Yet despite this extra expenditure on the part of caring adults, "choice," as a central tenet of children's fledgling citizenship, remains a problem. Oliver suggests, not unreasonably, that there is an illogic to the way in which "choice" appears to be an organizing principle in an elementary school cafeteria. The cafeteria becomes a site of tension between, on the one hand, a commitment to the notion that individual choice is of the utmost importance and, on the other hand, the worry that choice itself might be inappropriate—not only because the kids are so young but also because freedom of choice in the cafeteria might itself be specious since so many of the available choices are unhealthful.

The narrative of *JOFR* cannot sort out the illogic of blaming children's bad choice making for the failures of the school lunch system, and thus abandons the question. Rather, the high school appears to be a more accommodating site for emphasizing the possibilities of caring citizenship as a solution to poor nutrition and obesity. These students are ostensibly old enough to be expected to exercise their food-freedom responsibly. Thus, in the high school, choice re-emerges as an organizing principle of the Food Revolution. Here, Oliver does not impose punitive measures. Rather, he addresses an assembly of students, urging them to care enough to join him. He calls upon the students to freely make the choice to "help" him. "I can't fight you," he says. "I shouldn't fight you. I don't want to fight you. I need to get you on board. So, today at lunch—I haven't taken away your french fries, there's the good news, folks!" Oliver goes on, "and if enough of you choose my dish, then I'm going to get permission to roll this out through the other schools. And if you don't, I won't. Basically it's all up to you. What happens in this country, this food revolution, comes down to this school, on this day, with you students." He "put all the responsibility on them" (*JOFR* 1.5 2010). Oliver offers students the *opportunity* to choose his healthful lunch line over the cafeteria's three typical lines serving fries, burgers and pizza. This is a test of their ability to exercise their freedom to choose responsibly. Further, if they fail to demonstrate care in their cafeteria choices, they will be responsible for perpetuating the poor nutrition of school lunches, and by extension, according to the show, the broader problems of ill-health and obesity. Thankfully, for the sake of narrative resolution,

Oliver is pleased to discover "all the feet" have voluntarily found their way to his food line. Although the possibilities of choice and the specious condition of "freedom" in the cafeteria were undermined in the elementary school, the show reestablishes these in the high school where the contradictions are less obvious. The episode thus re-positions the responsible exercise of individual freedom through caring about food as the key to improving nutrition in school lunches and, in turn, to making a healthier Huntington.

## Caring Corporations, Private Monies, and Public Support

Ostensibly, *JOFR* is able to smooth over (or ignore) the challenges to free choice that emerge in the schools. However, within the narrative—no matter how much individuals care about the school lunches—their caring efforts are not in themselves sufficient to overcome a dearth of funding. One might assume that such problems call for legislative changes (and even, perhaps, that caring citizens could organize to *demand* such changes). On the contrary, on *JOFR* no demands are made of the State or even of private industry. Rather, consistent with broader neoliberal trends, the show argues that the existing state monies must be maximized through creative re-organization and that additional funding ought to be secured through private appeals to powerful individuals and corporations. It is in these appeals that caring citizenship is put into practice. The show thus not only demonstrates how caring citizenship ought to be performed, it also opens up space for private industry and powerful individuals to position themselves as caring as well. In responding to citizens' appeals, powerful entities can *choose* to care.

By choosing to care, the powerful can distance themselves from their structural role in school lunches and cast a caring mantle over their profit-oriented (rather than people- or health-oriented) activities. For example, Oliver visits US Foods,[2] the distribution company that, in his words, "suppl[ies] the schools with all this processed stuff." Oliver acknowledges that, "Frankly, they make a lot of money doing it." Yet the show does not suggest that the fact that US Foods profits so handsomely from contracts with the USDA and public schools is either good or bad. It is simply and neutrally true. The profit motive does not preclude Oliver, acting privately on behalf of the revolution, from appealing to US Foods as a potentially caring corporation. He explains, "if I want to get all the schools in this area cooking fresh food, from scratch, then I do need to get these guys on side." Just as Oliver's revolution requires the caring participation of parents, cafeteria cooks, teachers and students, it likewise needs US Foods—a company whose only responsibility is to generate profits for its shareholders—to similarly care and "want to help" (*JOFR* 1.5 2010).

---

2   US Foods a privately held corporation and is the second largest food distribution company in the US, fourth largest internationally.

Yet, while parents and teachers (and even children) risked *failing* to care about food and nutrition, and thus being "part of the problem," US Foods is positioned only as a resource. As a distributor, US Foods is figured only as a tool to be used by choice-making actors in a free market. The choice-makers, the show argues, are ultimately responsible for the quality of the food that ends up in the cafeterias. The interests of US Foods are erased from the narrative. For example, a US Foods representative tells Oliver, "we have pretty much anything our customers want.... We let them decide." In addition to the processed stuff, the rep shows Oliver rooms full of what Oliver describes as "fresh meats, fresh vegetables ... local produce ... loads of stuff that's really good, but the problem is that US Foods has box after box of the processed foods because the schools are buying junk." The problem, Oliver argues, is that the schools demand junk, therefore US Foods sells it to them. The structural reasons that cause the schools to buy junk—its cheapness, the contracts industry has with the public school system, the distribution of public money—are not discussed as sites at which change is possible. Neither is the solution to regulate systems of food pricing, supply or distribution differently. Rather, according to the show, the schools themselves (and by extension, the student-eaters within them) are responsible for the bad food, and thus they must change their demands (*JOFR* 1.5 2010).

Despite that US Foods (which, as noted, profits from the poor nutrition of school lunches) is apparently immune from failing to care, demonstrating care is within its purview. Specifically, Oliver needs the helping corporation on his side so he can "stay on budget." US Foods becomes a cooperative player in the food revolution.[3] It does not offer funds toward the revolution, but appears in a supporting role by helping Oliver maximize the way existing monies are spent to purchase different products. The way in which US Foods assumes this supporting role reinforces the notion that innovative thinking on the part of caring individuals—that is, the creative reorganization of the apparently inadequate resources—can reasonably stand in for policy changes that would provide additional resources.

This kind of creative purchasing, however, is not offered as a complete solution. For Oliver, this solution hooks up with a broader project to generate monies for the revolution. Yet, Oliver does not seek public funds that would be put into place through budgetary policy changes. Rather he seeks private donations that can be secured through individual, entrepreneurial efforts. The show does not suggest that the State's inadequate funding of the public school lunch system is a problem. Instead, it argues that State support is not required, for financial needs can be met by free-market actors who respond to the pleas of deserving individuals. For example, the show creates narrative drama around whether Oliver can sufficiently "impress" Doug Shiels, a representative from Huntington's privately-owned

---

3   Interestingly, US Foods is singled out as a potentially exemplary corporation. Branded "junk food" products in the cafeteria, however, are blurred out. When the narrative arc calls for the vilification of particular foods, the brands do not appear. When the notion of a "helping" corporation furthers the narrative, the brand is celebrated.

Cabell Hospital, to secure the funding required to expand the school lunch program throughout Huntington and make it "sustainable" after the show leaves town (*JOFR* 1.5 2010). Shiels, however, is not the CEO of the hospital. On the contrary, he is the vice president of the hospital's public relations department. His job is to make the private hospital and the "business community" of Huntington look good on national television. I do not point this out to argue that Shiels represents is a "false" form of caring or that he is an ersatz stand-in for a "real" leader within the hospital. Rather, the fact that Shiels—and the private hospital, for that matter— appears as an obvious and indispensible partner for the "revolution" demonstrates the extent to which the private sector, profitability and branding (in this case, the Huntington brand as one that will attract successful businesses) are inextricable from the performance of caring at present. The funding promised by this form of caring is offered not only as an *acceptable* substitute for structural change, but as a *desirable* one.

Although the show does not give sustained attention to the contradictions of calling on the private sector to meet the needs of school children (for it would undermine *JOFR*'s broader argument), it cannot help but gesture to them. A caring corporation is not obligated (by law, for example) to contribute funds to public projects. Rather, it contributes *voluntarily*. If the company does not expect to benefit from the project, it may not invest. As far as the hospital is concerned, Shiels suggests, *JOFR* ought to improve Huntington's image to make it more attractive to businesses, investors, and employees. The hospital considers withholding the funds because of the negative publicity *JOFR* generated about Huntington by calling it the "Unhealthiest City in America" and making a spectacle of residents' fat bodies. The private sector can grant and withhold funds as it serves shareholders, economic growth, and its brand. Without intending to do so, *JOFR* thus demonstrates the limitations and precariousness of private sector solutions.

Significantly, however, *JOFR* does not suggest that private institutions are the only possibility for securing funds for the "Food Revolution." The show also includes appeals to the public sector. Yet these appeals are consistent with the broader argument of the show, specifically, that powerful actors can be persuaded to care (that is, to voluntarily invest) if caring individuals (parents, students, Oliver, for example) publically demonstrate the degree to which they care about food and nutrition. For these individuals, public displays of willingness to perform unpaid labor and extra effort (*not* public protest or collective demands), is the means through which caring is put into practice. For example, Oliver's "gang" of teenage "ambassadors" takes on extra (unpaid) work to garner support for the school lunch program. Under Oliver's tutelage, they prepare a gourmet meal in a large, fancy restaurant for local bigwigs, including a state senator (*JOFR* 1.3 2010). The dinner is an especially illustrative case for understanding the role of the public sector in *JOFR*. After dinner, the teens come out of the kitchen to applause and tell heartfelt, and tearful stories of lives affected by obesity. This labor-intensive performance of the teens' own caring investment elicits the desired response: the senator agrees to "support" the revolution. *JOFR* neglects to give viewers any details about the content of this "support," however. The fact of the support is more important than its content. Indeed, when one of the teens pipes up, "you guys make the change and we'll believe it!" thereby stepping out of caring-citizen character, Oliver appears in the following frame and chalks the statement up to the boy's caring investment in the project and a misunderstanding on the boy's part, for the senator had already offered support (*JOFR* 1.3 2010). Thus, emotions like anger and indignation are deemed inappropriate, though in this case, endearingly so. With respect to the senator, the show is unconcerned with the line at which his individual performance of caring (or "supporting") ends and his public service as a political representative for the people of Huntington begins. If, as the show argues, caring is itself transformative, there is no need to distinguish between the two. The absence of details also leaves the role of public funding entirely vague within the narrative (though there is some indication that the funding is nominal, for the bulk of the money, the "sustainability" of the revolution depends upon a commitment from the hospital, not the public sector). The show thus disarticulates the senator's support from the State and from policy changes. His support is merely another demonstration of individual care.

The show thus argues that self-styled pleas for assistance—made persuasive through the demonstration of individual motivation, skills, and aspirations, not collective demands or structural changes—are the key to improving school lunches. Although the public sector appears in the show, the fact that *JOFR* represents a plea to state actors like the senator does not mean that the show adequately addresses political and economic questions connected to the poor funding of public school lunches. Rather, the entrepreneurial logic on display at the banquet is consistent with the broader argument of the show: that change is effected, not through collective demands, but through the private empowerment of caring citizens.

## Bureaucratic Red Tape

Despite considerable inconsistencies, *JOFR* manages to stitch cafeteria workers, parents, children, private industry, and the Senator into a narrative of personal responsibility. However, the show's final site of intervention, the USDA rules and regulations governing school lunches, is not so easily incorporated. It is here that the tensions between the show's commitment to the possibilities of caring citizenship and the inadequacy of this solution become most visible. For Oliver, fact that these rules and regulations favor foods lacking in nutrition becomes a major obstacle. This might appear to invite a structural critique: one could, perhaps, historicize how the USDA regulations came to be, explain what they aim to do, and analyze the interests that keep them in place. One might also expect that Oliver's collision with the USDA rules and regulations might point to the limitations of neoliberal interventions, that is, the limits of individually caring citizenship in the face of massive structural obstacles. However, instead of taking it seriously,

*JOFR* redefines the problem of USDA rules and regulations as bureaucratic "red tape," abstracting them from their historical, economic and political context, and returning the narrative to the caring citizen. For Oliver, complicated rules and regulations get in the way of caring work. For example, he sits down with Rhonda McCoy, food service director for Cabell County schools to learn how to meet the USDA rules for his meal plan. Oliver begins by drawing a comically simplistic diagram of his goals on a dry-erase board: an arrow connecting a plate of "hot, fresh food" to the "little mouth" of an elementary school student. In stark contrast to Oliver's seemingly "obvious" drawing, McCoy pulls fat binders, books, cards and other documents out of her bag and stacks them on the table. They contain the USDA rules, guidelines and recipes for school lunches. "What does USDA mean?... Are they your boss?" Oliver asks. Interrupting McCoy, Oliver complains, "This is very academic stuff... I just wanted to cook some food and this is like a-a- math test. There's so much red tape there. It's so complex!" Oliver's diagram sprouts a tangle of additional arrows and becomes so spotted with circled, technical terms it is incomprehensible. "I don't understand it. Normal people in America won't understand it. Without question, I am more confused than ever!" (*JOFR* 1.6. 2010). Here, "red tape" is a problem less because it prevents meeting the nutritional needs of children, than because it thwarts the wishes and work of the caring citizen.

Throughout the series, Oliver dismisses USDA guidelines as a "headache" (Ouellette and Hay 2008, p. 60). For instance, though Oliver repeatedly attempts to have the pink and brown, sugar-filled milk removed from the cafeteria, it inexplicably reappears. French fries count as a vegetable, while Oliver's "seven-veg" pasta sauce does not measure up. The guidelines, he decides, "are a load of rubbish" (*JOFR* 1.6.2010). The utter irrationality of the "red tape" places it beneath analytical attention. Its ostensible irrationality is mobilized to invite viewers to be outraged. This "outrage," in turn, is aimed at encouraging citizens to empower themselves to care more: if they are parents, they can take steps to improve nutrition at home; if they are cafeteria staff, they can put in "more effort" at work; if they are students, they can choose wisely in the cafeteria and learn about cooking and nutrition.

Yet, although the show appears to dismiss the rules and regulations as an abstract outrage, because the school remains a public one, State bureaucracy cannot be circumvented by the actions and investment of individuals. Dealing with the State is unavoidable. Hence, while the broad narrative of the show argues that the solution to ill-health in Huntington depends on the individual empowerment of caring citizens, the final episode cannot sustain this logic seamlessly. This episode explicitly points out political and economic impediments—impediments that cannot be surmounted by individual caring citizens. Despite having convinced parents, students, cafeteria staff, and teachers to care, to put in extra effort, and to make better individual choices, Oliver discovers that the school has continued to place orders for processed food and instituted "processed food Friday" to use up the surplus. When Oliver confronts McCoy, she cites pre-existing USDA contracts with food companies. Outraged, Oliver speaks directly to the camera: "The USDA has got to evolve and change, and support communities that want to change. If you've got everyone in the world who wants to cook food from fresh, but they can't buy the food from fresh? That's a problem." Here he acknowledges that caring and wanting change are not sufficient to producing change. He makes an explicit moral judgment about the presence of "junk food" in the schools stating, "it's not okay that the government at large allow this." He does not argue that the USDA must be circumvented through the use of private monies and caring individuals, but rather that the USDA itself must change (*JOFR* 1.6. 2010).

The USDA change Oliver calls for is not a matter of making exceptions for individually deserving schools. The series ends as Oliver muses, "I mean, maybe I can use my influence to ask the USDA to make special allowances... But," Oliver shrugs, "maybe the USDA needs to make special allowances for everyone" (*JOFR* 1.6, 2010). Here the show calls into question the prevailing notion that individual demonstration of "specialness" is the criterion by which individuals become deserving of things that otherwise might be understood as rights (access to healthful and affordable food, or fair wages with which to buy food, for example). The suggestion that "everyone" might deserve "special allowances" fits more comfortably with a rights-based discourse than one, like neoliberalism, which takes individual responsibility and entrepreneurialism as central tenets.

## Conclusion

*JOFR* works to fit the problem of school lunches and ill-health in Huntington into a narrative of personal responsibility that obligates individual caring as its solution. It extends this obligation to waged workers in the cafeteria, to parents, to children, and to individual corporate and State actors. The show celebrates the possibilities of individual caring citizens, who exercise choice responsibly. It applauds the benevolent "helping" corporation, which, in demonstrations of largess, might offer resources to help caring individuals circumvent a state represented as a "headache." Yet at each step, this logic threatens to fall apart, revealing the limitations of the neoliberal solutions to which the show maintains a commitment. In the cafeteria workers' complaints, in the demonstrated precariousness of private funding sources, in children's unwillingness to choose appropriately, in the absence of healthful choices in a purportedly "free" market, and in Oliver's own concerns about the absence of adequate health coverage in the United States and the injustice of bad school lunches, *JOFR* opens up a critique of the status quo not entirely resolved by its narrative. As Oliver and others on the show visibly bump up against economic, political and bureaucratic constraints, *JOFR* suggests that a broader intervention is necessary if sustainable change in school lunch programs is to be effected. This is an intervention that is not achieved solely through the cultivation of individually caring citizens, the cooperation of food distribution corporations, and the enlistment of private funders. This suggestion may open up possibilities for *JOFR*'s particular brand of caring citizenship to link up with broader collective efforts to demand legislative changes. However, I would stop short of celebrating the contradictions within the show. Although I do believe they open up possibilities, they do no go sufficiently far. *JOFR* suggests that caring citizenship is itself a kind of nutrition intervention. If we aim to "do nutrition differently," the structural obstacles to good health—like poverty, low wages, limited access to healthcare, USDA rules and regulations in which french fries count as a vegetable, inadequate funding for school lunches, and the contractual relationships between the USDA and food distribution corporations—must not only be mentioned as abstract outrages, but rather be given sustained attention and analysis. Doing nutrition differently requires an understanding of how nutrition interventions, like *JOFR*, not only work to change the food options in school cafeterias, but also offer narrow definitions of health problems that delimit the kinds of solutions that are thinkable.

## References

Binkley, S. (2006). The Perilous Freedoms of Consumption: Toward a Theory of the Conduct of Consumer Conduct. *Journal for Cultural Research*, 10 (4), 343-362.

Brown, W. (2005). *Edgework: Critical Essays on Knowledge and Politics.* Princeton, NJ, USA: Princeton University Press.

Cruikshank, B. (1996). Revolutions within: self-government and self-esteem. In A. Barry, T. Osborne, & N. Rose (eds), *Foucault and Political Reason: Liberalsim, Neo-liberalism and Rationalities of Government* (pp. 231-251). Chicago, IL, USA: University of Chicago Press.

Guthman, J. (2011). *Weighing In: Obesity, Food Justice, and the Limits of Capitalism.* Berkeley, California, United States of America: University of California Press.

Harvey, D. (2005). *A Brief History of Neoliberalism.* New York, NY, USA: Oxford University Press.

Jamie Oliver's Food Revolution. (2010, April 29). *Broadcast.* ABC.
Season 1; Episode 1. (2010, March 21). *Jamie Oliver's Food Revolution.* ABC.
Season 1; Episode 2. (2010, March 26). *Jamie Oliver's Food Revolution.* ABC.
Season 1; Episode 3. (2010, April 2). *Jamie Oliver's Food Revolution.* ABC.
Season 1; Episode 4. (2010, April 9). *Jamie Oliver's Food Revolution.* ABC.
Season 1; Episode 5. (2010, April 16). *Jamie Oliver's Food Revolution.* ABC.
Season 1; Episode 6. (2010, April 23). *Jamie Oliver's Food Revolution.* ABC.

Ouellette, L., and Hay, J. (2008). *Better Living through Reality TV: Television and Post-Welfare Citizenship.* Malden, MA: Blackwell.

Stobbe, M. (2008). *Videos you may be interested in W. Virginia town brushes off CDC's poorest health ranking.* Retrieved January 23, 2012 from USA Today,

November 16: http://www.usatoday.com/news/health/2008-11-16-wva-poorest-health_N.htm.

# Concluding Questions

This volume has been collected in order to generate dialogue about doing nutrition differently. The volume has not tried to cover "all bases," nor do we pretend to have provided a comprehensive starting point for thinking critically about nutrition. Our hope is that the questions and concerns, the omissions and contradictions that emerge out of the volume will be seen as fodder for new conversations. The contributions of the volume are varied and hopeful, serious and playful. We inaugurate much more than we complete; we activate rather than finalize.

We began the volume by expressing a collective interest in moving beyond what has come to be seen as a narrow and repressive approach to diet and nutrition. We referred to this narrow approach as 'hegemonic nutrition' and described it with the adjectives: standardized, reductionist, decontextualized and hierarchical. By way of a conclusion we end not by summarizing what has already been said, but by articulating some of the questions that the volume yields. These are questions that try to move towards a nutrition that exceeds universal metrics, that expands rather than reduces, that pays attention to context, and that refuses to succumb to the hierarchization of knowledge.

1. What would it take for the *partialness* and *situatedness* of nutrition science to be recognized? What are the processes through which such a recognition could emerge (in higher education, professional practice, policy circles and the public eye). Through what means could mainstream nutrition science be put into deeper dialogue with other, diverse food and health knowledges/practices?
2. Who is complicit in the ongoing dominance of Western nutrition science? Researchers? Academics? Practitioners? Media? By what means can we identify the nodes and networks through which Western nutrition science becomes hegemonic, and through what means can the dynamics of these nodes and networks be shifted in the name of doing nutrition differently?
3. What is hegemonic nutrition? How can we identify it? How can we better theorize it? What does it look like on the ground? Where are its 'boundaries'? Where are its soft points? What are its *affects*; what does it *do*?
4. Thinking about the metaphor of the "iceberg of nutrition"—what is it that remains 'hidden' beneath the tip of hegemonic nutrition? How can that which is hidden be seen, heard, felt or understood? Another way of putting this is: what is nutrition when we move beyond hegemony?

5. Is an *ontology of nutritional difference* desirable? What does this mean for thinking differently about the human body? How is nutritional difference different from nutritional relativism?

6. If we are to broaden the scope of what is included within the bounds of 'nutrition' then who should we look to in order to solve some of our current dilemmas? What new calls to action emerge from this broadening of nutrition and to whom are they calling? For instance, should nutritional professionals start to address poverty? If so, what kinds of preparations are necessary from a pedagogical standpoint? Should nutrition become a more interdisciplinary field?

7. What changes would be needed for nutrition to be understood as an issue beyond the individual? Can and should nutrition move beyond the model of changing individual behavior? How can those interested in doing nutrition differently be empowered to understand work on nutrition as a larger structural problem? Whose responsibility is it to work against the structures that prevent certain people and groups from nourishing their own and others bodies?

8. What is the role of public health offices and policies in doing nutrition differently? Specifically, what should be the role of public health professionals in demonstrating the deep and complex connection between food systems and other social and ecological systems? What other roles, shifted or created, might also need to be filled?

9. How can scholars, activists and professionals work to recognize the ways in which nutrition interventions tend to reproduce white, middle-class privilege (e.g. in their emphasis on individual responsibility and self-sufficiency at home), without being stymied by this recognition? In other words, what would it mean to do nutrition *intervention* differently? Would a different approach entail a different starting point and/or a different scale? If so, what are the challenges and potential cautions of these new points and scales?

10. What is the role of pleasure in doing nutrition differently? More broadly, how and to what extent can bodily/bodied knowledge help to inform the ways in which we sustain our bodies through food? What does the body know about nutrition? Along similar lines, what role does emotion (always embodied) play in nutrition? In what ways could emotion be harnessed or worked upon in an attempt to do nutrition differently? How does emotion extend beyond the individual body that eats?

11. Can the notion of decolonization—and more broadly paying attention to colonial legacies and discourses—assist an expansion of the project of nutrition along non-hierarchical and non-racist lines? Salient here may be not only matters of what counts as 'good food' and 'expert knowledge' but also who is seen as in need of intervention and upon whom research is conducted.

12. Is it possible to express nutritional positives and negatives without participating in problematic processes of 'othering'? Is it possible to learn from various traditions and cultures without forcefully appropriating nutritional knowledges that are beyond one's own heritage or scopes of reference? What is the role of education in doing nutrition differently if we are to be cognizant of the need to move beyond a hegemonic emphasis on education-as-cure.

13. What is the relationship between forms of creative expression (like poetry) and nutrition? What kinds of creative expression could enable new understandings of bodily sustenance?

14. Where do and should ecological concerns enter into nutritional dialogue? Does thinking freshly about nutritional knowledge production also help us to think freshly about the production of ecological knowledge and vice versa? How can we better understand the political ecology of nutrition?

15. What work can be done to push against the demand for expert nutritional knowledge, which tends to push scientists and others to elevate and prioritize the latest nutritional theories? In a similar vein, can and should nutrition attempt to move beyond quantitative biomarkers and other standardizable metrics for bodily health? What is the role of qualitative and quantitative understandings here?

16. How can new technologies for collecting and visualizing spatial and/or body-centered data help to continue the project of doing nutrition differently while resisting attempts to standardize the body-food relationship? Can such technologies assist nutrition scholars, activists and professionals in contextualizing nutrition rather than further abstracting it?

17. What is the relationship between hegemonic nutrition and neoliberalism? On what scales do neoliberal policies and practices affect the doing of nutrition? Does a focus on neoliberalism invite sustained attention and analysis of the structural obstacles to 'good' health?

18. What do historical analyses of hegemony in nutrition provide to the project of doing nutrition differently? What kinds of insights and affects emerge through scholarly analysis and/or emotional engagement with historical narratives?

19. Who should take the lead in doing nutrition differently? What changes to the power dynamics between activists, community organizers, scholars and policy makers need to occur? Is it possible to occupy nutrition? What would that look like? Should we? Dare we?

20. When we do nutrition differently, what else starts to change? What does this mean for body-food, body-land, and body-community relationships? What does an ontology of nutritional difference imply for our systems of food and health/wellbeing? What shifts might begin to be demanded in social and ecological systems more broadly?

# Index